BIOLOGY OF KNOWLEDGE

BIOLOGY OF KNOWLEDGE

The Evolutionary Basis of Reason

by
RUPERT RIEDL

Zoological Institute, University of Vienna, Austria

with the collaboration of

Robert Kaspar

Zoological Institute, University of Vienna, Austria

Translated from the 3rd German Edition by

Paul Foulkes

A Wiley-Interscience Publication

JOHN WILEY & SONS

Chichester . New York . Brisbane . Toronto . Singapore

COPYRIGHT

British Library Cataloguing in Publication Data:

Riedl, Rupert
Biology of Knowledge
1. Knowledge, Theory of
2. Cognition
3. Perception

I. Title II. Kaspar, Robert III. Biologie der Erkenntnis. *English*

121'.3 BD181

ISBN 0 471 10309 8

Printed in Great Britain

CONTENTS

FOREWORD

Our account of how organisms have evolved phylogenetically presupposes that every successful adaptive step marks a gain of information about what furthers them in their environment. To evolve is to gain knowledge, where "knowledge" is not the technical term of the philosophers but the ordinary word, as when living systems, by gradually adapting, have come to bring out the laws of nature, somewhat in the way the eye has reconstituted the laws of optics. This biological approach frees the study of cognition from the shackles of philosophical enquiry: the phenomenon is no longer confined to reason but becomes itself an evolving object.

Our position thus differs radically from philosophical epistemology, for we examine the basis of reason not merely from its internal principles, but from a comparative phylogenetic study of all cognitive processes. Thus what we study is no longer "identical" with the subject that gathers knowledge, but lies mainly outside it, while the method remains that of comparative natural science, avoiding the limitation arising when reason must establish itself on its own.

The aim of our investigation is to find out under what conditions those mechanisms have developed which we must suppose to be functional preconditions for human reason to arise, what functions they contain and how they become further differentiated. The totality of these preconscious cognitive attainments is called the 'ratiomorphic apparatus', better known as ordinary unreflective common sense. On the one hand it is a precondition of all rational reflection, on the other its cognitive modes can be developed from its phylogenesis as a whole. We thus obtain an overview of those principles that help all living things to gain knowledge of this world.

The history of our enquiry is not very old. Freud and Jung were the first to study the unconscious. The subject was extended by Piaget, Brunswik and Chomsky and developed into a science by Konrad Lorenz. Owing to him it influenced psychology in America and then, through Donald Campbell, Karl Popper's philosophy back in Europe, where this development led Gerhard Vollmer to set up a first framework for "evolutionary epistemology"; we now expect this to be filled out with empirically testable hypotheses. That is our special task here.

The solutions, offered by this "evolutionary epistemology" are thus grounded in empirical testability. They encompass the old problems of the a priori, of inductive generalisations, of reality, in particular those of certainty, comparison, causality and finality, and in sum the trilemma of the epistemological grounding of reason. Moreover, this method shows that any cognitive mechanism works correctly only in the environment for which it was selected; beyond these limits, it quite misleads us in our search for new knowledge.

The limits of our cognitive power readily follow from that. We shall show how unreflective and reflective reason together can overcome the barriers beyond which they lead us into error. Since in our current setting the limits of that carefully adapted guidance are greatly exceeded, it may be useful to have an objective sorting of the reasonable and the unreasonable in our thinking.

My thanks go to my family and friends for their patience, especially to Konrad Lorenz and Erhard Oeser and the faithful members of our Altenberg seminars. Without Lorenz, evolutionary epistemology would not have arisen, without Oeser it would have remained uncorrelated with the scientific theories of today. Again I thank my dear wife for sympathetic illustrations, my pupil, Robert Kaspar, for taking charge of the very tedious parts of our work, and Dr. Annemarie Illsinger for carefully checking the text. Once more I thank the publishers, Paul Parey, particularly Dr. Friedrich and Dr. Rudolf Georgi for the splendid promotion of this book.

Vienna, Summer, 1979 Rupert Riedl

Our thanks must now go to our readers too, for only a few months after the publication of the first edition, we have had to prepare a third revised edition. The rapid turnover of the first two shows that we have circumspect readers, who understand that a deeper insight into the structure of reason can profit us all.

Vienna, Summer, 1980 Rupert Riedl

PREFACE TO THE ENGLISH EDITION

German speaking readers have welcomed this book with some pleasure and surprise. This is both pleasant and surprising, for what here seems a brand-new solution to the dilemma of human reason, was stated forty years ago by Konrad Lorenz, in his study on the Kantian a priori. At the time, it went unnoticed; Copernicus and Darwin, too, refrained from fighting for recognition. As usual, it is the next generation that will not stand for the new wisdom and its blessings being with held from the public.

In the sciences, opinions had indeed been divided. Biologists found the theory fascinating but hardly new. Inorganic scientists did find it very new but not so fascinating. Those who till now have found 'evolutionary epistemology' most instructive are students of education, historians, sociologists and economists. It was not so much the theory but its outcome that began to interest teachers, managers, economic advisers, and even politicians (who valued it as confidential briefing). In Eastern Europe, it was decided to "enlist the theory in the Leninist compact between Marxist philosophy and modern natural science."

As regards philosophy in the West, the case stands otherwise. To begin with, could speculative epistemology be turned into a science of cognitive gain at all? Neo-Platonists dislike our way with the rationalise controversy, Neo-Kantians our solution to the origin of the a priori, and positivists our admission of induction, heuristics and hermeneutics. Some philosophers imagine that they would lose ground, others mistake me for one of their number.

The Church, oddly enough, seemed friendly, given that well-meaning philosophers and psychologists had hailed the theory as a third Copernican revolution: some misgivings might have been expected within the clergy. However, we are beginning to learn from history; today, we find it a matter of education whether we reconcile the views of spirit as formed by evolution or by creation.

Now for the English reader! Our dealings so far have been with speakers of German, a language allied with philosophy that has but lately cut its philosophic apron-strings, not without some alarm. In contrast, English goes rather with the world of things. Moreover, for historical reasons, epistemology in Austria was always closer to the English variety than to the German.

This should make for lively discussion. May success attend the efforts of my publishers (Parey in Germany, John Wiley in England) and the diligent care bestowed on the translation by Dr. Paul Foulkes.

Vienna, January 1983 — Rupert J M Riedl

INTRODUCTION

"Life itself is a process of acquiring knowledge"
Konrad Lorenz(1)

A reader intending to discover something about the basis of his own rationality will enter some provisos. Before becoming entangled in a thicket of facts and arguments, he will want to know what to expect, the more justifiably so when current text books wander through an amazing hall of mirrors, distorting our view of a world that has itself grown very strange.

The author, therefore, declares himself: he is a biologist by profession and, as is well known, biologists are people who try to discover new facts about the structures and processes of life by applying scientific methods. His collaborator is his pupil; and yet this possessive pronoun rather oversimplifies the case in at least one particular. For the author is not a philosopher, a view shared, if not by all biologists, certainly by all his philosopher friends, and that is what matters. This may be due to the widespread view that we can hardly tell what philosophy really is(2), but we should want to share with it the "love of wisdom" its name implies.

Thus we are not willing to abandon the scientific method, for "in science, in contrast with philosophy, only those theories survive that stand the test of experience"(3). This is important; for some of the consequences to be drawn from our investigations are incisive. That they are objectively testable must be part of the ethos with which we approach these matters. Some consequences will go beyond traditional biology. Indeed, the foundation of biology itself already lies outside that framework, namely, in its scientific methodology; and this lies within the means of our cognitive capacity. In short, biology and knowledge will yield a unitary method.

The biological acquisition of knowledge

The present "Biology of Knowledge" is intended to show not merely how biological knowledge is acquired, which one hopes is dealt with by "introductions to biology", but rather how the biological cognitive process itself works, how organisms become aware of their life problems, which algorithms (methods of calculation) have proved reliable for dealing with information from their surroundings and activities, and how they become anchored in the organisms. This attempt readily suggests itself, for we can always test and observe the inherited mechanisms which border on wisdom in helping the organisms to solve their problems. Behaviour theory wholly relies on this topic, unless indeed it avoids such complex events. We shall meet these achievements as the innate mechanisms of possible experience(4).

The process of evolution

The first thing to investigate is what development led to these mechanisms. Again, this is a biological question, and concerns the process of evolution. The theorem of evolution has indeed by now encroached on the areas of chemistry, psychology, language, the humanities, technology and epistemology(5); but, as in the past, one relies on biology. For since the time of Lamarck, Lyell and Darwin(6), the theory of evolution in biology has by now withstood the test of two and a half centuries. One might, of course, object that the theory of evolution takes us to a theoretical level. Though true, this is only a matter of convention, since the theory has long been so probable as to border on certainty. The evolutionary process is theoretical in that the processes of micro- or intraspecific evolution are amenable to experiment, but those of macro-evolution (trans-specific evolution,(7)) only to observation; just as we can test the theory of gravitation in terrestrial mechanics, but only observe in celestial mechanics. However, few would worry because we cannot test experimentally whether the sun will rise tomorrow, although there have been such people(8), and they might well have seen this as untestable.

The evolution of cognitive mechanisms

Even in the third presupposition of our investigation, in which we combine the first two, we do not yet meet the untestable. This is the evolutionary theory of cognitive mechanisms, the core of our objective. The theory draws its facts from several independent sources.

For a start, there is biological behaviourism, which has revealed the gradual building up of mechanisms, the function of which is to apply to organisms successful programmes for deciding in the face of more and more complex situations and happenings in their environment. "Life itself," concluded Konrad Lorenz, "is a process of acquiring knowledge"(9). Thus our eyes reproduce the laws of optics. Moreover, we find that in data processing and stimulus conduction, the higher solving procedures presuppose lower ones and thus continue the latter's algorithms as well. We can be quite brief here because much will have to be drawn from this source later(10).

Secondly, the systematic conditions of evolution offer a further basis. Since data, whether processed in animal or human brains, lead to the same solution patterns (normative, interdependent, hierarchical and transmissive), an old puzzle revives: do not our thought patterns determine those by which we describe nature? Are we not projecting on nature our own way of grasping order, because we cannot think otherwise? Having discovered the system conditions of evolution, we can explain why, in all its structures, the "animate order"(11) takes on these patterns, which are incomparably older than the methods for observing and processing them: therefore, the patterns of nature must determine those of thought(12). Selection had to pick out the most appropriate results. The order of the real world must be presupposed if we are to learn from it.

Thirdly, the very continuum of evolution supports our thesis. Today, this field of research ranges from the evolution of molecules right up to that of civilisation.

Even in the prebiotic era the strategy consists in seizing chance events and preserving the structural laws resulting from them, as shown by Manfred Eigen(13). This "order-on-order" principle, as Erwin Schrödinger(14) foresaw, extends throughout the whole evolution of organisms; and, according to Piaget(15), it is continued in the development of the child, in adult behaviour (Lorenz and Eibl-Eibesfeldt)(16), in the preconditions of language (Chomsky and Lenneberg)(17) and in the phenomenon of handing down cultural patterns (Otto Koenig)(18). The unity of this "strategy of genesis"(19) is now well established.

Fourthly, the continuum of the cognitive process itself underpins this unity. What Freud and Jung anticipated(20), namely the background of preconscious processing, or what the psychologist Egon Brunswik called(21) the quasirational "ratiomorphic apparatus", becomes scientifically comprehensible. The basic preconditions of reason thus prove to be innate. Again, Konrad Lorenz was the first to see this(22); and soon after the corresponding evolutionary mechanisms were clarified, in the psychology of the cognitive process(23) by Donald Campbell, in the processes of theory formation(24) by Karl Popper and in the development of the sciences themselves(25) by Erhard Oeser. The whole process of the evolution of knowledge, therefore, is accessible to science. This is now material for evolutionary epistemology.

Among its postulates is the view that our conscious cognitive powers are the most recent superstructure in a continuum of cognitive processes as old as life on this planet; besides, as the most recent stratum in the knowledge gaining process, it has so far been least tested and refined against the real world; that this reason must encounter fundamental difficulties, because of the rapid growth in what can be grasped and reflected on, together with equally swift changes in conditions of testing and selection; and that the conscious cognitive powers in their rational basis must be thought of as a further development in their phylogeny, whereby - and this is crucial - an investigation of their development and limits and of these very difficulties would become possible.

We further expect that this concept may develop into a complete theory and Donald Campbell has made an initial attempt at this(26). Gerhard Vollmer is the first to have designed the structure, for an "evolutionary epistemology" with a view to the whole. We refer to this work explicitly for without it we could not now proceed as if we were concerned with a long established discipline. Vollmer(27) says "its framework should be filled out by a precise theory. This requires above all the working out of a system of categories for human experience supported by biological and psychological evidence, the severing of objective from subjective cognitive structures, a more precise definition of the term 'partial isomorphism' (i.e. a matter of the degree of agreement between natural and experienced patterns)(28), the formulation of empirically testable hypotheses on innate cognitive structures and of hypotheses on their phylogenetic development." Evidently this task is so obvious that I have pursued it at the same time as Vollmer. It was the subject of my seminars, of the scripts of Robert Kaspar's(29) and of my lectures during the last few semesters. From these activities, the present book has grown.

The solution of some puzzles of reason

The aim of this book, therefore, lies in the solution of some puzzles of reason. In the midst of evolving cognitive mechanisms we wish to secure a standpoint from which we can discuss reason rather objectively. We should like to solve some as yet unsolved basic epistemological questions precisely within the framework of the theory of their evolution. For the expert, these are simply the problems of reality, inductive inference, our attitude to causality, space and time, Kant's a priori of pure reason and the a priori of the aims of our judgment. For the non-expert we shall explain them carefully. These open questions have not only made science insecure and brought structural research in biology to the edge of ruin: by their consequences, they have divided our world view, from antiquity to the present day. They have rendered the basis of our knowledge, and hence ourselves, susceptible to manipulation.

These consequences too we shall pursue. We shall demonstrate that reason and experience, idea and reality, mind and matter, have been separated unjustly and to our detriment. We shall expose reason's pitfalls, and those who, with their reason, urge these pitfalls against all reason; against humane values and against man.

CHAPTER 1
BIOLOGY AND COGNITION

"It is the greatest scandal of philosophy that, whilst nature around us — and not only nature — is being destroyed, philosophers continue to debate whether this world exists."
Karl Popper

"With every new scientific discovery and every further philosophical treatise on induction, the statement of the philosopher C D Broad seems increasingly confirmed: Induction is the triumph of natural science and the disgrace of philosophy."
Wolfgang Stegmüller(1)

We all know what "biology" means, but "cognition" seems to belong to the philosophers; besides, what could biology have to do with philosophy? Have not the sciences, as the offspring of philosophy, long since divided what is scientific in their inheritance? Have they not all gone into the real world, abandoning the philosophers in their sacred groves of Platonic Ideas? Lamarck still called his system of zoology a "philosophie zoologique"(2), but that era is past.

At one time, philosophers were ready to die rather than recant: Socrates preferred the cup of poison, and Giordano Bruno the stake; this, too, is no longer fashionable. With a wink, Will Durant summed it up thus: "Nothing remains for philosophers, save frozen heights of metaphysics, the childish guesswork of epistemology and purely academic disputes over an ethic that has lost all hold on men"(3). What flourishes are the exact sciences, and most of their textbooks no longer mention philosophical problems.

What however, unites us with philosophy, even the most exact of exact scientists amongst us emancipated offspring, is the conviction that we are dealing with the real things in this world; the postulate that nature is objective. Otherwise, we could have saved our efforts and remained in the pleasure gardens of unchecked flights of fancy. Yet we are united in the ambitious view that we are gradually moving towards the truth(4), namely the agreement of object and experience, which, of course, needs testing. What is idle is to quarrel over who is entitled to do this: for example, the biologist or the philosopher. One thing alone counts: what method of testing leads to what results and with what degree of certainty.

Who should learn from whom

As biologists, we therefore ask how it can be and how we must understand that we come to know our objective nature ever more objectively. This concerns the relation between the knower and the thing to be apprehended, between the possibilities and the objects of knowing. In this, biology and epistemology are linked. Pointless to argue about who is to learn from whom: forebears from their offspring, or the reverse, as commanded by the respect our elders teach us for their own wisdom.

Modern biology, like the biological interests of children, began quite rightly with collecting and classifying. It has gone on to ask what causes living processes and states. Today it extends from the molecule, through the origin of molecular information and the "knowing" exhibited by instinctive regulators to the formation of our consciousness. The most general question it asks is how we are to grasp that larger systems of molecules, such as those that make up the reader or the author, can organise themselves to the point that, in their own view, they themselves come to think about molecules.

Modern epistemology begins with John Locke and this task: "to enquire into the original, certainty, and extent of human knowledge, together with the grounds and degrees of belief, opinion and assent"(5). Its earlier metaphysical background recedes more and more and, in its further development, critical traits become reinforced, a positivistic tendency(6) sets in and finally an evolutionary one. Its most comprehensive question becomes how we must take a relation between knowledge and real things, since knowing that presupposes knowing each of the two, and so on indefinitely.

If, then, epistemology asks how we can gain knowledge about knowledge, biology asks how knowledge can arise out of itself, and that, at any rate, is what epistemology has to do with biology today.

THE DILEMMA OF REASON

The puzzle of knowing about knowing, the source of reason's reason, may seem a children's riddle, especially to one whose daily concerns (scientific ones included), what we call "the serious business of life", prevent him from seeing life as a whole. His own success, judged by the successes in his special field, proves to him how safely he can ignore the solution of this riddle. However, should he then have to explore the foundations of no more than his own speciality, he will find that ultimately there are none. Therefore he will either have to admit that knowledge is without foundation, or he himself must begin with the puzzle.

However, one feature of this riddle is that it must be as old as our thinking about the process of coming to know, and that, leaving aside unjustifiable prejudices, it has so far remained insoluble. One formulation goes back to the didactic poem of the pre-Socratic philosopher Parmenides(7), since when its basic form has not altered. Subject and object remain incompatible with their opposing aspects of thinking and being, ideas and reality, mind and matter. What has changed is merely the question as to which of the two contains the basis for the other, and therefore the home of truth.

Subject versus object

Whatever we can know about the objects of this world is necessarily built on our experiences as the subjects; and likewise for our volitions, thoughts and actions. Subject and object appear both as opposites and as foundations for all knowledge (Fig. 1). Contradictions arise at once. Subjectum is the experiencing, opinion-forming, thinking and willing being; literally, that which is "thrown underneath", in the sense of an ultimate ground on which everything rests. Objectum, on the other hand, means that which is "thrown against" the subject. However, the vagueness of these terms is at once emphasised, says Konrad Lorenz, "in that they have exchanged their meanings since the time of the Scholastics"(8) so that in English, "subject" is often used throughout in the German sense of the word "object". And even in German these terms are odd(9).

Now, it is asserted, we cannot know what objects of this world are really like. An apple is neither red nor sweet if there is nobody to see and taste it. It then contains only particular molecules and reflects a particular wavelength of the spectrum. What is most certain, the fundamentum, in all knowledge must therefore be the subject, and in it, its thought; even the certainty that anything exists at all resides in my thinking. Thus, Descartes(10) proclaimed Cogito ergo sum, I think therefore I am. However, when can I be certain of my most certain knowledge, or rather, whose certainty is the most certain? The picture in the mind is too inconstant. A drop of wine and the mind will embrace the universe; or on some other occasion, withdraw miserably into a single hollow tooth. Yet if I myself am uncertain, then whose certainty would be binding? The wise man's or the fool's? In that case certainty would last only so long as no-one contradicted; neither expectation, nor our senses nor our neighbour. However, they all contradict: we trust neither our hopes, nor our senses, let alone our neighbours. That is why we call the certainties experienced by a subject "subjective", which ordinarily means "partial, prejudiced and linked with chance evaluations". So the confusion becomes complete, since we derive the adjective "objective", meaning real, unprejudiced, and from objects that we can know only subjectively and in a prejudiced manner. Descartes' dictum in no way helps us out of this dilemma. We can even turn it round: sum ergo cogito, for only "because I am, I think". In any case it is a vicious circle. Neither subject nor object carries the ground of certainty. Reason has simply split our world.

Idea versus reality

Let us, therefore, try more radical remedies and cut the knot. Plato rejected the Sophists' view that the senses are the touchstone of truth and man the measure of all things. For in that case any fool or dreamer could give as good a picture of the world as any other. What the "noisy crowd of the senses" tells us is a Herakleitean flow of change which by itself could never lead us to perfect truth. What is perfect is only the idea, or eidos, of a thing. Every individual man dies, what lasts is the concept of man. Every real triangle is imperfect and decays. It is only the idea of the laws of triangles that are perfect and eternal.

This opens the abyss of the perishable against sense experience, which at best shares in the ideas. No sense experience of nature can go beyond founded opinion, so that all scientific discourse must be myth. The bonds of testability are thrown off, and metaphysics arises. The "Idea" marks Western destiny: with Augustine it becomes the thought of God, with Schelling the soul of things and with Hegel absolute truth(11). Thus reason has split the world into idea and matter, developed idealist philosophy and prepared the ground for the incompatibility of ideologies.

Idealism versus materialism

Idealism, or more appropriately idea-ism, explains the world by its highest laws, ultimate causes or ultimate purposes. Aristotle's commentators selected from his ingenious theory of the fourfold causes in the world(12) the so-called final cause which, as if from the future, purposefully draws events to their ultimate goal of perfection. Thus, what normally suits our understanding of human action and culture, now claims the status of a universal explanatory principle and leads to intractable contradictions. Man is created for spirit, life for man and matter for life. The thousand million year round of evolution was danced for our human purposes. What presumption! How this contradicts the roots of our history and the continuing torment of creatures! Besides, there is the problem of teleology: the final cause would have to act from the future back into the present, which is incompatible with what we know about causes. Moreover, we can no longer distinguish between idea and natural law. The type, as unity in the patterns of organisms has been called since the time of Goethe(13), becomes an idea; and the nature of the natural system of organisms, a pattern of thought. Already Kant had called it a "philosophical scandal" that it had so far failed to give a "proof of reality to the external world"(14), but Solipsism(15) showed that only the notion of an individiual (say, of the reader) exists, and that his assertion that all that appears around him exists only in his imagination cannot be disproved by reason. "In my opinion", so we quote Karl Popper today, "it is the greatest scandal of philosophy that, whilst nature around us — and not only nature — is being destroyed, philosophers continue to debate — sometimes cleverly, somtimes not — whether this world exists"(16).

If with the primacy of ideas we cannot break out of the vicious circle of uncertainty, then perhaps with some opposing materialist philosophy we may. This too goes back to the pre-Socratics, and begins with the Renaissance, with the development of science, with Galileo and Newton, and with the sole admission of efficient causes from the four Aristotelian ones. Thus, exact science begins to explain the world exclusively from the powers detected in matter. Finality becomes the declared enemy and efficient causality is used as a weapon against it. A further split occurs in attempts to explain the world: reductionism arises, according to which thought processes are attributable to physiological, and in turn to molecular biological processes, and these again to physical and chemical ones. The mind does not exist, or else it is no more than a complex material reaction; while reason must govern the processes of genetic molecules(17) of life and of thought. Once again, what presumption! How this contradicts the as yet elusive complexity of living organisms! As Karl Marx affirmed, Hegel's universal explanation, which countered a half-truth by turning it round and institutionalising the incompatability of ideologies, is simply stood on its head. This means simply that what two half-truths cannot explain is delegated for the final determination of "genuine truth" to ideologists, who can then impose it on their peoples as the basis of all certainty. The vicious circle of reason closes once more.

Determinism versus indeterminism

Amongst the minor upsets that have meant strife between idealism and materialism let us examine one more. Idealism, indeed, tends towards proving God's existence and being determinist. If one admits ultimate purposes towards which the world is striving, then it must be predestined to reach them. Scholasticism even recognises exemplary causes, an ultimate divine purpose. From this we can infer a pre-established harmony in which everything is basically meaningful. This makes individual freedom problematic or at least rather puzzling, witness the Church Fathers' attendant efforts(18).

The natural sciences also began in a determinist style. Man became a machine(19) and physicists imagined a Laplacean mind that knew the movement of all particles and could thereby foresee the whole future. Only microphysics discovered the loop-hole in causal determinism(20) and showed that this physical chance could extend into the macro-region(21). Thus, materialism was inclined not only to deny God's existence but also to assert indeterminism. The world became the product of change and, since chance is the opposite of plan and order, there was no harmony in this world. Thus, freedom seemed guaranteed even for man, but since he arose wholly from chance mutations, he was devoid of sense.

So it was that Teilhard de Chardin(22) saw a deep purpose even in wars, whilst his contemporary fellow-countryman, Jacques Monod, brought scientific proof of man's senselessness(23). In this conflict of reason, therefore, even sense and freedom begin to exclude each other.

Reason versus experience

Because of these contradictions of reason, pure idealists and materialists, indeterminists and determinists have become rarer, and people have remembered the "actual" epistemological process, which must be an interaction of reason and experience. The roots go back to ancient philosophy. For, "paradoxically, the problem of truth arises only with the discovery of reason. Knowledge, the Sophists said, comes from the senses. From the baboon's or from the wise man's, Plato asked? Thus knowledge must come from reason of which Aristotle formulated the laws. Then how do you know, Pyrrhon asked, that the wise man is wise? Therefore, said Epicurus, back to the Sophists. Next, the sceptics asked what good will that do?" We can trust neither experience nor intelligence. "Nothing is certain, Pyrrhon concluded; and when he died, his pupils, who loved him, did not grieve for him, since they could not be sure that he was dead"(24).

Rationalism versus empiricism

When Greece and Rome faded away, the positions hardened. The Scholastics chose reason and founded rationalism; science since Galileo and Bacon chose experience and empiricism arose. So, in principle, things remain, as does the lack of trust in either half-foundation of those contradictory epistemological structures.

This had already been derided by Parmenides: "Most mortals have nothing in their fallible intellects that did not come in through their fallible senses." However, modern discussion(25) begins with David Hume's critical question whether anything that we believe can be justified with adequate reasons, and if so, what. The central problem is then revealed, namely, that of space and time as forms of intuition, of induction of causality, and of the a priori.

A Priori versus a posteriori

Immanuel Kant(26), in his critical writings, examined the limits of reason and judgment, and isolated the prior assumptions that cannot come from experience, because without them no experience is possible. These are the a priori features of reason and judgment. Kant's findings are helpful only because they sharpen the issue. He states very clearly what constitutes reason's dilemma, but that does not solve the problem. For the a priori (literally "from what is prior") cannot itself be questioned. The chain of assumptions is endless, which shows precisely that reason cannot establish its own grounds.

Why should we accept probability, which pretends to give us foresight (even if only roughly) about foresight that we cannot have? How many white swans must we see and yet err in concluding that all of them are? Who makes us trust that more means true? Bertrand Russell wryly suggested that a madman who thought he was a scrambled egg could be dismissed only because he is in a minority(27). Here we have the problem of a priori probability, as well as the Hume-Kant-Popper problem of induction, on which all inference from the special to the general depends. All inductive science, the whole of natural science, will stand or fall with the justification of such inference. Not even modern inductive logic offers certainty or sufficient reason for it, as Rudolf Carnap and Wolfgang

Stegmüller have shown(28). As for Karl Popper, he declares it to be self-contradictory(29).

The riddle of induction

The problem next stands out in the question what entitles us to take similarity for sameness or to expect the same causes for the same effects, since comparability and causality ultimately come not from experience but must be presupposed if we are to learn from it. If empiricist philosophy in turn can show that the presuppositions of reason lie not in experience but beyond reason itself, where is that place? If this "beyond" lies in questions concerning a cosmic ground, the meaning of being and happening, what came before birth and human awakening(30), then, by definition, we are back with metaphysics. The circle, as ever, has closed, only more clearly.

This dilemma of reason spans two and a half millennia, the whole history of our civilisation. Our account is thus short and rough. Nor can we here examine the ramifications of the problem. The reader should consult works like that of Erhard Oeser, who treats the theory of science as a reconstruction of its history, reaching quite novel syntheses(31). Here we merely show that the dilemma of reason is as old and widespread as the gradual discovery of our own reason, with two results.

Thinkers divide

The first result is a division of thinkers. What to the practical man might hitherto have seemed mere hair-splitting and a bent for depth, has vital consequences. The peculiar structure of our reason has disrupted our picture of the world at a point decisive for our own self-understanding, where mind and matter touch. Konrad Lorenz has made this very clear(32). Subject versus object, idea versus reality, reason versus experience have led to idealism versus materialism, determinism versus indeterminism and rationalism versus empiricism. All these incompatibilities are facets of the same fracture, which soon become institutionalised in faculties and anchored in law. Discussion between manifestly mindless natural sciences and unnatural humanities was broken off and two half-truths have set.

One might think that this is just the old academic chicanery — until one becomes aware that we are ruled by it. Social psychology pretends to show that reality is a social construct and every society determines what is true in its
reality is a social construct(33) and every society determines what is true in its presuppose in order to know the world, is precisely least certain, the more that particular reality must be raised to the status of genuine reality. This happens through those secret bonds between the general prejudices of the time and the special ones of the particular political demands. In the cycle of the unknowable and the necessarily certain, insecure societies leave it to their upper reaches to establish principles since at the top somebody must know what in their view

should be their sacred rights. Ideology becomes a substitute for the foundation of truth. Since two contradictory half-truths must lead to two rightful claims to the whole truth, they are incompatible and have long since divided the world. This is the vital consequence.

The trilemma of knowledge

The second result is our disappointment that reason's dilemma must lie in reason's own peculiar structure. Clearly, we can never find the ultimate presupposition, because we must presuppose that each presupposition must itself have a further one. Epistemologists have discovered that this unsolved dilemma can take three forms. Hans Albert speaks appropriately of a trilemma of knowledge(35). Either we must recognise circular inference or an infinite regress of preconditions, or simply stop arguing about the problem. This is where matters stand today.

The wearied reader will welcome a stop to the debate, as does the author. At the same time, our own discussion is only just beginning. This is due to a second disappointment revealed by epistemology: it shows us that the problem of reason cannot be solved from within pure reason itself. This very despair gives hope to the biologist, for his position enables him to establish reason from without. This is evolutionary epistemology.

Traditional epistemology had indeed expected rescue from the fact of man's successful gain of knowledge. Let us defer judgment. However, as biologists, we can prove that the living organism, with its gain of knowledge, has successfully pictured its world for more than three thousand million years. For, in Popper's words, "if we assume, and I do, that our search for knowledge has so far been successful and we now know something about this world, this success is unintelligible, unlikely and hence unaccountable; an appeal to an endless sequence of unlikely chance events is no explanation. (The best that we can do, I think, is to examine the all but incredible course in which these chance events have developed from the formation of the elements to the formation of organisms)"(36). So we shall.

IS LIFE MORE RATIONAL THAN REASON?

Mere common sense already tells us that this cannot be: reason is simply man's privilege. We act rationally, animals do not. Moreover, this same common sense leads us to judge that a person's behaviour is human or animal. If somebody who is starving steals, we call this behaviour only too human, although every animal does likewise. If someone kills his family and then himself, we call it bestial, although no beast ever behaves like this. Clearly, we must be more careful. Remember what the crowd laughs at in a zoo: only such animal behaviour as caricatures our own. Nor do we blame ourselves for projecting our own bad features onto our neighbour, the more readily to condemn him. However, since we are talking about reason, we need firmer ground.

What, then, is reason? The concept has indeed changed with time, just as our respect for reason has grown with the growth of reason's claims. In Old and Middle High German, the term "Vernunft" denotes the "activity of apprehending, sense perception, comprehension, insight"(37). The Scholastics distinguished between sensatio, ratio and intellectus, roughly corresponding to perception, concept framing and formation of ideas. Eckart used "Vernunft" as translating ratio, Kant as translating intellectus. Today the term means reflectiveness, insight, spirit, intelligence(38). In short, we are no longer concerned with correct perception or reception of messages, but with correct and apposite reaction to it.

What is rational in reason

What makes reason rational is being right, appropriately digesting and reacting to correct data. Reason must be equal to the task in hand, avoiding the snares and pitfalls on the way to successful solutions if any exist. What is here meant by "tasks" and "success"? The task will ultimately concern life, from choosing a suitable pair of shoes to adopting an appropriate design for living. In another sense, these tasks can range from deliberate deception to self-sacrifice, so long as success is assumed. Whenever we fail, whether in choosing footwear or life plans, in deception or sacrifice, in retrospect, we call our action unreasonable. Success is therefore to be gauged by solving the task. The ultimate goal is to achieve a purpose(39): that is how success operates.

Rationality, therefore, lies in success in life, in a positive balance between success and failure. We all know that it is irrational to concentrate only on books or only on sport (however successfully on each), for we suspect that this might diminish success in life as a whole. Success in life is furthered by whatever may promise gain in security, health and strength, foresight, knowledge and wisdom, prudence, influence and power in order to make one's own life more secure and prosperous, as well as what may depend on it, the life of the group, society, mankind and the biosphere. It concerns individual, species and environment.

The inanimate displays no reason: it seems neither rational that tectonics erects mountains, nor irrational that erosion carries them away again. However, we do wish to see reason in what we ourselves make. As we shall see, the boundaries of reason are like those of purpose. Everything alive seems to be full of reason. How rational, or perhaps appropriate, that storks migrate south, that beavers dam the water and cuckoos carefully lay their eggs in strange nests! However, are we not simply projecting our own reason onto the behaviour of animals?

The rationality of processes that diminish entropy

To judge this objectively we must view the matter from outside plants, animals and man, preferably from physics. All living processes decrease entropy. Increase of entropy is the universal tendency in nature to go from order to disorder; as do the objects on our writing desks, except for our constant tidying up (Erwin Schrödinger)(40). Living things, on the contrary, generate order where none existed before. Being open systems, they circumvent the entropy principle in that they must expel more disorder than they can build order within themselves. This ordering is the same in protoplasm as in full organisms, as Manfred Eigen has shown(41). A source of chance events continually produces small changes in memory and environment of these systems, sorting out the less fit by reducing identical replication (that is, their increase), while allowing the fittest to multiply more rapidly. What maintains itself, or survives, undergoes changes made optimal by trial and error, so that the rate and kind of change suit its structures(42). Moreover, the survivor develops the features most suitable given its environment. What, then, is meant by "suitable features"?

Some have indeed regarded Darwin's(43) selection principle as tautological: "the survival of the fittest" as no more than "the survival of the survivors". However, this view is wrong, which matters greatly for our conclusions. For in a given environment we can objectively predict what will be abler or fitter. The more suitable or fitter system is always the one with features best adapted to the prevailing regularities. The survival or permanence of living systems must lead by trial and error to a progressive extraction or reproduction of the natural laws surrounding them.

Why the eye is radiant

"Life itself is a process of cognition" as Konrad Lorenz put it(44). What matters in this gain in knowledge is not so much an urge for truth but, trivially and pragmatically, immediate success in life, an ever more positive balance between success and failure, what we experience as reason or suitability. The one limit this trivial pragmatism sets is that only what turns out to be essential and proves itself should be stored as knowledge. That is the rationality of the matter; it can lead to the greatest heights and certainties of knowledge.

That is why the eye is radiant, as Goethe(45) foresaw. Otherwise it could not see. Today we know that evolutionary mechanisms have led it to exhibit all the relevant natural laws of optics: the lens and its movement, the diaphragm and its control of aperture, the focal plane and screening, all these are developed as in the best optical instruments with rational precision (Fig. 2).

We are therefore justified in trying "to explain the rational plan, that creative Nature (once called Providence) pursues with mankind"(46), as Kant has hinted. Where, then, lies the greater rationality and success in life: in preconscious or in conscious reason? Many have asked this question: Rousseau(47) favoured the first, Kant the second. However, as Kant allows, "all natural abilities of a creature are meant to develop fully and suitably one day"(48). All living structures depend on grasping these natural laws essential for survival and on adapting themselves accordingly. For Schrödinger(49), to live is to consume order; for Lorenz(50), information. This holds for every structure, from bodily shape to component parts down to the position of molecules, from the simplest to the most complex behavioural structures. The environmental regularities that decide success in life are reproduced by trial and error, encoded in the genetic material and refashioned in spatio-temporal structures controlled by genetic building and operating instructions: witness the perfect hydrodynamic shape of the dolphin, the mechanically optimal structure of bones (Fig. 3), and the osmotic efficiency of cell membranes.

A selection of rational world pictures

All living structures then contain stored knowledge, a kind of judgment on the laws under which they exist. Given identical replication, this prejudges regularities that the next generation will face. Given that individual conditions of life will be repeated, this moreover means that pre-formed judgments and expectations persist. This stands out well in the way behaviour repeats itself in time. A selection of rational world pictures arises, each consisting of a system of suitable prejudgments on the relevant portion of the real world.

Konrad Lorenz has fully described the evolution of these rational world views as the development of the "backside of the mirror"(51). Here we must confine ourselves to the principle: layer by layer, as stimulus conduction, nervous system, brain, sense organs and cerebrum have developed, ever expanding genetic programmes grasp, store and suitably reproduce the regularities of ever wider portions of this world.

Amongst the simplest programmes are kinesis reactions, as in protozoa with fixed body ends. A single receptor suffices which, to slow down or speed up if during undirected motion in water droplet, it receives adverse or favourable messages (Fig. 4). The animal will therefore rationally stay mainly in a prolific area, like a mushroom collector. From this point onwards, avoidances and arrangements develop further(52). A new form of control arises with the complex sense organs. These are stimulus filters, innate release mechanisms: they determine which messages will rationally lead to which reactions. Children, parents, or enemies should be detectible from far off and in advance. Therefore, evolution provides children, parents and even partners with signals so as to ensure that the receiver decides correctly (Fig. 20). Instinctive actions and imprinting follow(53). The highest form of what the genetic material learns in the way of operating instructions shows itself in ordered systems of instinctive movements (Fig. 37). Even in the stickleback, the sequence of releases starting from the habitat first concerns the occupying of hunting grounds then the choice between fighting, pairing, nesting, brood care and, in conflicts only, between impressing, biting and pursuing. The logic of programmes becomes all-embracing.

What is rational in these genetically acquired judgments about the environment, what is rational in these "world pictures", is that they consist entirely of what is practical and pertinent: the pictures must be completely right for the domain for which they were selected; nothing impractical, remote, untested or fictitious could be admitted into them. For only direct and continued testing of the result ensures that what can be experienced is built into the molecular memory of the genetic material.

The hurdle of the first learning phase

How fast and how correctly the genetic material can learn constitutes an initial difficulty. It shows itself in the accuracy of forecasts based on extrapolating from what has been learnt. Biologically, it is the ratio between adaptive speed and selective control.

In this first evolution, learning is incredibly slow: it takes at least a million years to establish an inheritable trait in a species of higher animals(54). The testing is thus very thorough: transient conditions are not absorbed. The problem is then the speed of learning, since extrapolations always lag behind the more rapidly changing environment and, as will be shown(55), retain excessive deficiencies. Evolution must push towards greater learning speed: given sufficiently powerful brains, this was achieved through the invention of individual learning.

Individual learning begins when closed programmes are opened, with conditioned reflexes. Training already shows that a programme that rationally causes the salivary glands to flow at the mere sight of food, can also be associated with the sound of a bell, provided it is rung often and regularly enough. Learning speed is thereby increased by seven to nine orders of magnitude, from a million years to days and hours, but control is correspondingly decreased. This new success is likewise furthered in higher animals, through curiosity and play, particularly in young animals. However, remarkably little is learnt(56) from the learning success of a neighbour. Only the individual animal pays for mistakes of extrapolation, and when it dies its individual learning disappears with it: for the species, success and risk remain small.

This changes only with language, when man begins to appear. With language using concepts and even more with writing, a new code system arises and once again experience gained is secured for the species. This is the second evolution, which presupposes the first and continues its principles.

The hurdle of the second learning phase

This, too, has to do with learning speed and testing, namely inadequate or delayed testing of extrapolations derivable from what is learnt. New insights spread through the populations like wildfire. Moreover, growing consciousness further incites extrapolations. Through consciousness, we can extrapolate in an imaginary space, and keep on extrapolating to higher levels as far as fancy will take us; initially without let or hindrance, witness the flood of weird and unreal combinations in dreams and fantasy, if we capture them accurately.

The second evolution builds entirely on the structures of the first, indeed more strictly so than used to be thought possible. About this detail more later. The principle of the evolutionary mechanisms continues unchanged. All creative learning, which alone continues the evolutionary process, remains subject to contrary chance and necessity. In the brain, the new elements enter chance combinations, selection retains what has proved reliable, and this in turn is codified in the memory of civilisations, an imperishable inheritance for the individual populations. Even the acquiring of knowledge in science seems to go this way, if one compares the discussions on this subject by Paul Feyerabend, Thomas Kuhn, Erhard Oeser and Karl Popper(57).

The magician's apprentice of evolution

However, being accelerated, the process has changed and man has become the magician's apprentice of this evolution. Innate programmes and regulators, which give wise direction, are built over. Intellect is placed above instinct, and reason above common sense. The natural inhibiting mechanisms, like those that the first evolution incorporated in all aggressive weapon-bearing species(58), are circumvented by the development of long-range weapons. Man's liability to imprinting can be manipulated now by advertising, now by ideologies. His sense for collecting and activity is swept along by the exponential growth of the economy and of power. With his sense for group and community he constantly finds himself fenced in behind new commanders.

Man's effect on his own environment is no less accelerated. The testing function, which alone divided the suitable from the senseless, he now claims for himself. The chance dispositions of his power decree now this, now that, what counts as good and what as objectionable. He must now himself determine his significance and observes that his old regulators have abandoned him. He has to decide what his world ought to look like, and sees that he is destroying it. The magician's apprentice begins to be alone with his reason.

The unreasonable in reason

Testing not only lags behind the speed at which imagination constantly creates new chance combinations, but man adjusts the tests in such a way that our fancy is confirmed. We always create a set of assumptions as self-evident and of prohibitions that allow an unchecked growth of superstition. By contrast, no animal's programme of world pictures can contain completely senseless elements: they may become fallible only at the boundaries of the environment for which they were selected. As Lorenz put it, "to believe pure nonsense is a privilege of man"(59). Wherever man has abolished testing, such nonsense can flourish unchecked. He divides mind from matter, making mind into machines and matter into minds. He mistrusts what he constantly recognises and entrusts himself to what he cannot know. He doubts that the world is real and at the same time ruins it. This is what is unreasonable in reason.

Evolution indeed goes on testing, even whether the species Homo sapiens is suitable. Our suspending it could only be temporary. Besides, we know that man "errs in his thinking, often for centuries until he adapts himself by experience" (Friedrich Dessauer)(60). The cost of such errors falls on many generations and often it is our own fictitious constructions that provoke the controversy and then the struggle between powers. The strife is rarely settled by reason, but rather by fire and brimstone. The chance alliance of new powers must then decree the new true truth. All this belongs to the irrational in reason.

Let us, however, admit at once that conscious reason is by far the youngest product of this evolution, and therefore doubtless the least tried. However, since the second evolution has made the results of this reason inheritable we all partake in its risks. The whole tribe must answer for any collective nonsense; on that ground we may personally complain.

Is, then, the preconscious reason of living creatures more rational than our consciousness? On this opinions differ; Rousseau against Kant, and Voltaire against Leibniz(61). It is simply one of the puzzles of man; it is indeed difficult to be objective. If, however, the rationality of reason exists objectively on the positive balance of life success, in that it raises the chances of survival of the species by protecting its individuals and living space, then perhaps our conscious reason may be able to learn from the preconscious kind. For, on the one hand, three million years of life success are in its favour, locked away in two million successful different species, and on the other, there are the uncertainties of the comparatively tiny life spans of a few incompatible civilisations.

Even if nothing were right in this comparison, nobody doubts that it can be useful to learn from life itself.

A STRATIFICATION OF HYPOTHESES

How, then, do the living solve the problem of knowledge, of acquiring sufficiently reliable insight in this world? Here, the conscious part of our cognitive process has entangled us in a dilemma of contradictions and incompatibilities, but life could show constant success; this difference may probably lie in the method.

Indeed, we found the sense of a cognitive process in the appropriate method for successfully solving the set tasks. These are for life its own tasks and success. As regards the method, however, so far we have established only that it must avoid the snares and pitfalls on the path to knowledge. This is our next theme, and we can foresee that it concerns methods of computing and deciding, special algorithms as they are called in mathematics and logic, for solving problems successfully(62). These algorithms will be expected to deal with the creative learning needed if the learner himself is to extract from nature regularities that he does not yet know.

This presupposes that there is something in nature to be learnt, an order in nature, for from disorder we can learn nothing. More precisely, order means law and application; mostly very frequent repetitions of the same(63). Once again, this is the reality problem where our conscious reason already runs the danger of deducing from nature the order that it has itself projected into nature; all the more so, as we can think or understand this order only in terms of law and application.

Life is a hypothetical realist

Indeed, as Bertrand Russell says, "There is no logical impossibility in the supposition that life is a dream, in which we ourselves create all the objects that come before us", a view that cannot strictly be refuted. "But although this is not logically impossible, there is no reason whatever to suppose that it is true"(64). There are many indications that support the reality of the world; Gerhard Vollmer has summarised them quite recently(65); but none of them is logically convincing. However, the solution that living creatures have found for the reality problem avoids deductive conclusions and depends on probabilities. If reality is neither to be proven nor refuted by deductive logic, then the rationality of hypotheses must help us ahead. As Donald Campbell and Konrad Lorenz(66) have discovered, life is simply a hypothetical realist.

In hypothetical realism, it is assumed, "that a real world exists, that it has certain structures, and that these are partly recognisable" and we will test "how far we can go with these hypotheses"(67). This is a very weak realism, but it provides a very firm basis. It supersedes naive, critical and strictly critical realism, for none of these is reliable in practice(68). Hypothetical realism, on the other hand, contains a method for improvement within itself, namely, the principle of self-organisation, a basic principle of evolution.

If we assume that we can learn from a hypothetical order of reality, then we must ask to what end. The object we know already; it deals with the solution of life problems with the aims of optimising life conditions. The process, however, involves the ability to foretell life problems, something in our cognitive world that corresponds to a judgment about future states, suitable judgments in advance, and hence correct prejudgments. This is a second principle of living organisms. For to learn something, however profound, without prospect of applying it successfully, cannot contribute to life success, whoever might believe otherwise today. What good would any knowledge be if there were no hope of deriving some benefit from it? Even feeling satisfied in the moment of discovering, say, that the neighbour is a fool but his wife a beauty, sees the satisfaction as related to the future; let us admit as much. Knowledge in itself may serve the philosophical pleasure in truth, but living organisms require constant success as well.

The necessity for judgment in advance

For the cognitive process of living organisms this is trivially true. For example, what should the dolphin's "learning" of the streamlined shape through inheritance mean, if it were not expected that conditions would remain the same and the structure would be an advantage to the progeny? How could the inheritance of a bird species develop a feeding signal inside the beak of the young (cf. Fig. 20) if one could not count on an inborn release mechanism in the same inheritance programme, now of the adult bird, remaining the same generation after generation? From the one inheritance, we must be able to count on the maintenance of the inheritance of another species. Otherwise, slave-keeping ants could not have arisen; because these make the ants of another species their life-long slaves, by being able to rely on the hatching animal to recognise, all its life, the individual that helped with hatching as a representative of its own kind. So they simply steal pupae of another species and help them during hatching(69).

The result of all biological learning is an unbroken network of prejudgments caused by endlessly repeated unchanging learning tasks set time and again by the same coach; by the life problems of the environment as well as by those of their own organisation.

For all that, these judgments remain essentially prejudgments. We deliberately use this term from our conscious world of judgments, for we quite rightly mistrust prejudgments. A judgment in advance can never be certainly correct. It is always an extrapolation and therefore never conclusive: for it infers from the known to the unknown. This is the Hume-Kant-Popper induction problem, as we saw. Again, the algorithm of the living builds not on apparent contradictions of our inductive logic, but on probability.

What recurs from case to case, may be confirmed in the next instance too. Although this cannot be certain nothing else more probable can be foreseen for the next case. The probability solution is as simple as that.

One important thing must be presupposed, namely the constancy of nature. It is an indispensable hypothesis for the algorithm of the living. Common sense would express it as "nature does not make jumps"(70).

The stratification of learning

A third basic principle of the living follows the 'order-on-order' principle, as Erwin Schrödinger called it(71); order can only be built on order. For the knowledge-gaining process, it means that new knowledge can most successfully be gained on the basis of old knowledge. Hence the stratification of learning. We therefore insist that tried experience must be heeded even where it is to be extended.

We have already touched on this matter in connection with the evolution of cognitive mechanisms and will pursue it further. Learning begins with the learning of building structures and at the deepest level is a learning of molecules. Indeed, according to Manfred Eigen's theory of the hypercycle, life begins where relatively short self-reproducing ribonucleic acid chains (the legislative, later gene material) are taken up protectively by the executive (proteins, the formation of which they initiate) into a superimposed cyclical process(72). From here onwards, all further processes and structures, from those in the cells to those in the eye, of the dolphin and of upright carriage, are learned by trial and error.

This knowledge, stored in the structures and functions of the body, is the basis of all kinds of behaviour extending the body functions, from the simplest kinesis reactions right up to the hierarchy of instincts (cf. Fig. 4 and 37). Likewise, they are the common basis for knowledge that is now laid down first in short-range and then in long-range sense organs (in the circuitry of unconditioned reflexes)(73), in inborn release mechanisms and in signals. Their joint knowledge is the prerequisite for individual learning which begins in the opening or coupling of unconditioned reactions to the conditioned reflexes. Even at this point, knowledge in structures and programmes is laid down, including an unsurveyable world of possible nonsense or failure.

It must be clearly understood that an evolution depending exclusively on creating by chance cannot afford to give much scope to blind luck: the chance of drawing the right lot, in any lottery, must be inversely proportional to the number of lots. This again is very simple. However, the consequences, which we are only beginning to understand, are very comprehensive. My outline of this "strategy of genesis" as a whole appears elsewhere(74).

With regard to the learning process, the "tabula rasa" view offers no explanation. It is absurd to assume that a new stimulus, a new task, without a great deal of prior knowledge of the programme will find the connection to suitable reaction and problem solving through blind trials. We can easily calculate that no life spans offer any prospect of appropriately hitting even a single association(76).

The inborn teachers

On the threshold of consciousness, therefore, all the knowledge of the organism that has already been laid down in structures, as well as an abundance of fixed programmes in the whole of the sense and nervous system, forms what Konrad Lorenz calls the inborn teachers. Moreover, he has convincingly shown(77) how consciousness arose, and the complicated space structure in the canopy of a tree top, binocular vision of our ancestors, the operation of the hands before our own eyes, individual learning in the differentiated group of primates, and how extending the care of the young has led to the development of consciousness. "The mind did not fall from heaven"(78). Just as no organism would learn except by being compelled by evolution to learn, so selection has managed to create consciousness. For it offers enormous advantages, namely, an ability to test an experiment in the "centrally represented space", in thought, without having to risk one's own skin whenever one makes a mistake.

It is self-evident that this thought cannot be free from the preconditions on which it is built. For a start, the sense organs give it access to only very limited areas; a continuum of electromagnetic waves is divided into qualities, here experienced as heat or cold, there as spectral colours. When we look from a bridge into a stream, how obstinately the inborn instructors compel us to "see" the larger area of the field of view (namely the water surface) as at rest, and the smaller one (the bridge pier) as in motion! How a drawing of two linked squares forces it spatially into a cube(79), and indeed alternately into its two possible perspectives, and in groups (as in Fig. 21) into a common perspective. We call figures "impossible" (like Fig. 5) although we have just drawn them(80). In everything we see forms and interpret them, whether they refute us or whether, as in constellations, there are no forms at all (Fig. 6).

The unteachable instruction

Everywhere we are taught by judgments in advance; often they are right. The result we attribute, wrongly as a rule, to our rational intelligence. Only where it judges wrongly do we become aware of this intelligence, and find ourselves perplexed because it refuses to be taught. For example, if we rotate a cube made out of wire before a mirror, and look at it with one eye so that the images coincide, then we cannot see them rotating in opposite directions. A change in direction always takes the other cube with it. The one that rotates wrongly must also be wrong in perspective, it "appears" to become soft and compensates for this error with a remarkable belly dance. This is almost a caricature of our mistakes in rational theory formation(81), where the simpler solution seems to us the more elegant and also the more correct; whilst we have to imagine quite unwarranted additions to appearance in order to compensate for the discrepancies in the theory.

The ratiomorphic apparatus

A whole system of such instructions steers and directs our rational thinking, as a rule surprisingly wisely and rationally. Egon Brunswik very aptly termed it a ratiomorphic apparatus(82), since it operates like intelligence throughout the preconscious, but has nothing to do with rational intelligence, which cannot even correct its mistakes. In some basic questions of cognition and of algorithms (that is, methods of problem solving) it might almost be said to have a point of view quite different from that which we receive at high school.

It contains judgments on probabilities which rationally we would contradict: it lets us expect, for example, that the six on the dice will be more likely to turn up if it has not done so for some time. We can compare forms and from this draw conclusions about forms not yet known, although we persuade ourselves that logically this cannot be justified. It allows us to expect some causal connection in all that is repeatedly in sequence, although, rationally, we often cannot tell what this should be. It often drives us very tangibly to seek the solution of a life problem in some quite definite direction without our being able to see any clear reason for this. It constantly leads us to assumptions, to testing or rejecting, to the barely considered solutions of small and minute resolutions and momentary decisions. It envelops and directs us with a system of hypotheses.

It directs our effort to understand so much that children, as we know from Noam Chomsky and Eric Lenneberg(83), need to learn not language, but almost only words. It directs the formation of our concepts and our social sense so that we are cultural beings by nature(84). It guides that which we call our quite unreflective but nevertheless sound common sense, that keeps us constantly moving with opinions and expectations; which rehearses for us, quite unrequested and at life's every turn, whatever might be the case and, of that, what might be otherwise. Often something that could scarcely be called a thought closes a small circle of repetitive arguments, and an act of will is needed to break out of it because it is pointless. Then again, without our asking, it creates surprising solutions which we greet as “aha experience” and take as an explanation of knowledge from somewhere or other. Often, in agreement with Carl Friedrich von Weizsäcker and Konrad Lorenz(85), we experience being aware of the solution to a problem even before we have discovered what it looks like and how to formulate it.

For and against sound common sense

This unreflective everyday understanding thus performs astonishingly. No one can claim that the useful solving of problems began only with the establishment of school philosphy or of formal logic. We know of high cultures that have developed without any science; without the whole dilemma of reason. Therefore back to nature, said Rousseau. No, reason is still to be developed, said Kant. Out of what, asked Karl Popper, the dilemma of reason, or everyday understanding? This is the for and against concerning sound common sense, now the problem of knowledge.

Is life then actually more rational than reason? Probably so, especially as it seems not to know reason’s dilemma. On the other hand, it might not possess any kind of certainty. It is impregnated with theory and never goes beyond mere hypotheses. Where then would our logic have secure support? Here, we follow Karl Popper’s solution, since our own investigation will confirm it: we shall begin with ordinary, everyday understanding. For “knowledge never begins at a zero level, but always with some background knowledge”. And “if an estimate were not absurd”, said Popper, “then I would say that 99.9% of the knowledge of an organism is inherited or inborn and only 0.1% is made up of changes in the inborn knowledge; moreover, I believe that the necessary adaptive ability is inborn”(86). On the other hand, what common sense contains of epistemology is “extremely naive and completely incorrect in all its forms” and even misleading, indeed dangerous, because “unconscious assumptions from it continue to exercise a destructive influence” on the cognitive process of the sciences “especially on the so-called behaviourists”(87). Background, solution and misdirection form the theme of our subsequent chapters.

A system of rational hypotheses

Moreover, every conscious cognitive process will show itself to be steeped in theories; full of hypotheses. In this, too, it resembles its biological background. Let us clarify this at once. Biological knowledge contains a system of rational hypotheses, advance judgments which, within the framework for which they were selected, guide us with the greatest wisdom; but at its boundaries they lead us into error, wholly and abjectly. The system consists of the four hypotheses in the next chapters; as they arose one after the other, in the evolution of oganisms and, presupposing each other, as they succeed each other in stages as algorithms for the solution of problems of survival. To investigate the cognitive process, we thus adopt a position outside our own learning process, namely a biologically objective description. We shall explore the same questions in each of the four hypotheses: (a) which problem has life to solve? (b) which method of solution did life develop? (c) how did this inborn instructor influence the formation of conscious methods? and (d) what is it in the environment changed by us humans that is sensible in this instruction, but also the source of the nonsense?

This is to show in what consists the problem of natural cognitive processes, by means of which biological reason devises its solutions, and how, from that, the dilemma of reason is explained. Moreover, we shall outline what is biologically unteachable with the attendant objective, hopes and cares in its train. From this extended knowledge of our own peculiarly structured reason, our "lazy reason" as Kant called it, we must no longer lose sight of these aspects.

It seems as if the science of reason was following a second time the path full of hope and sorrow, along an unbroken chain of hypotheses and collective error, which life itself by trial and selection has already followed for three billion years. As long as this chain does not break, we shall remain on the way, we hope, to a deeper and more humane knowledge of ourselves.

CHAPTER 2
THE HYPOTHESIS OF APPARENT TRUTH

"Most mortals have nothing in their erring understanding that did not come in through their erring senses." **Karl Popper** (after **Parmenides**)
". . . except the understanding itself".
Gottfried Wilhelm Leibniz(1)

Should we now accept as true what we perceive? Would untruth be avoided, as soon as no one contradicted; not a neighbour, not conscience, nor the senses? And, if these conflict, whom then could we trust? Me, or a majority of opinions; your opinion or ours? Or might it be reason, which can tell truth from all deception? Who then is to decide between your and our reason contradicting each other? In short,

Truth and Lies

of gods, demagogues, society or the imagination(2) dominate the whole world stage, and all the smaller stages of individual consciousness; in the midst of that gaudy scene made of revelation and conjuring, humbug and discord that are our history. Ever since we mortals have been able to write, we have written these things down; from the speculations of the epic of Gilgamesh(3) down to the speculative philosophy of modern times(4).

The one thing that does seem certain is that any further question as to states and events in this world will make sense only if we can assume that its object is at least fairly probably true. Whether we ask how things can be compared, what is their cause or even their meaning: the question "what is truth?" is the question of all questions, as Anatole France says; for what question does not depend on this? That is why the hypothesis of apparent truth precedes all others, the starting point for the cognitive process of the living and the engagement of consciousness; our study of cognition must begin with it.

WHEN THE POSSIBLE WOULD BE CERTAIN

Whatever feels alive aims above all at certainty, while helplessness causes the opposite; and yet we often allow that certainty of judgment indicates a lesser kind of mentality, and conversely, that Faust's insight "that we cannot know anything" or Socrates' "I know that I know nothing" suggest a higher wisdom(5).

Nevertheless, it can be taken as certain that without "knowledge", or without an adequate perception of its own conditions of life, no creature could survive.

If reason leaves us in uncertainty

What precisely do we know in knowing that without people to see electromagnetic waves, taste molecular structures and reflect on proportions, there would be no cognitive contents such as red, sweet and beautiful? How can creatures be attuned to the laws of their world when it must be quite different from what we perceive of it? If everything singly remains uncertain, how can the world as a whole become certain? If this world remains uncertain, might it not be wholly imaginary, as the solipsist maintains, a projection of the single one that does exist, namely the present reader? We know already that even our logic(6) cannot refute this view (solipsism).

This much we must admit. Suppose we wanted to refute a reader who held that his thinking was the only thing existing in the universe, then we might for instance say that he does not know the first word on page 186 of this book. However, if he has looked it up, he need only claim that this too was provided for in his thinking, which refutes our argument, and any other possible one. Here reason reaches the limits of its possibilities, in the ideal domain of extreme idealism (more correctly, idea-ism), among the Young Hegelians of the left, who include Feuerbach, Marx and Engels.

How, then, can reason render certain the assumptions that individuals make about the world, when it merely seems to him that dolphins swim, woodpeckers peck, solipsists write their books, there is water, trees, and people who can read(7)?

If common sense were not sound

Even if we rejected all this as fantasy, what would provide us with certainty? Bertrand Russell tells the story of an old lady who was taught by a solipsist that only her thoughts existed. She was so impressed by this that she told him there should be more thinkers like him in this world. Are we then guided only by "sound common sense"? Precisely the part of the understanding that is beyond erudition, the Cinderella of all boldness of intellect that cannot boast of any merit of the mind; unless it be modesty. We must of course admit that protection derives not from every intellectual efflorescence quite unconnected with sound common sense, but rather from that unreflective common sense which, almost unperceived and like a devoted guardian angel, guides us through the hundred and one small decisions we daily have to make all our lives.

Is truth, then, the conformity of the mind with things, the "adaequatio mentis et rei"? How could this come about, and which of the two is the measure of what is to be measured? Or is something true if it withstands testing, as William James(8) expressed it? Who, then, is to say what withstands the test? Was not the Ptolemaic view of the world, with the earth in the centre of the crystal bowl of heaven, reliable for centuries? May not today's well-tried views suffer the same fate? Are we not about to ruin our planet simply because we rely so much on such views? Even if the world were real, though we cannot precisely determine this reality, how can we be so certain and prescient of its events, in ways that constantly prove important for survival? What can ensure that today's friend will be friendly tomorrow, that our car will run, and even that the sun will rise again? Clearly, nothing! Moreover, those who despaired about this uncertainty are not minor thinkers(9).

If one does not know what chance is

Is it not strange that we cannot foresee how a die will move when thrown from a dice-box, although we have determined the geometry of both? It can obey only the laws of mass and acceleration, yet we can observe only chance. Besides, we do not even know what chance is, or even if it exists at all. We usually recognise it merely as a deficiency; the lack of certainty in prediction. Assuming there is a reality, science, following Heisenberg, finds real chance only in microphysical events in the world of quanta. Even there, its existence was denied by physicists like Einstein(10). If, as regards perception, chance seems merely a measure of our ignorance, how could we distinguish reliably enough between what we can and what we cannot foresee? Much of this reliability must be conducive to preservation, since no creature could survive if it took chance to be foreseeable but certainty not.

However, even if we do not despair over this constant proof of our uncertainty, but trust that practical experience in some unknown way will make us expect now chance now necessity, how could this world of half-certainties amount to a world that is certain? Even probability is shown by Kant to be assumption: reason presupposes it a priori but cannot establish it(11). Indeed, anything we have ever taken as fundamentally certain has in time been frustrated; as in our example where seeing any number of white swans and justifiably assuming that all swans are white, has merely led to the discovery of black swans(12). Many of our important expectations have turned out quite false: that the world is a flat disc at the centre, that species are immutable and atoms indivisible and that man has a special status. Why, then, do we expect that what we can imagine is roughly certain?

Although we can know nothing with certainty, not even whether the world is as it seems or whether it is real at all, what probability is, whether there is such a thing as chance, nor know how to derive certainty from mere possibilities, nevertheless, we are clearly here; we live and read, and somehow manage in this uncertain world; indeed all our ancestors over well-nigh 3000 million years of history must have mastered a much more uncertain world for them, otherwise we should not be here. Two million other species have done so too, for they still occupy the world along with us. They must all possess something of the truth, of sure judgment and of foresight about this world; otherwise they would not be with us, which shows that life has no need of deductive inference. It must cope differently.

THE PREJUDGMENT OF PROBABILITY EXPECTATION

It seems far-fetched that lower organisms like bacteria, amoebae or ciliates possess a kind of sure foresight over their world. In fact, they behave as if they did. This incredible circumstance results from the way creation works, namely by constantly narrowing, even if in turn only by trial and error, the scope of possible error. The censorship of selection, as we know, sorts out the successful individual from the chance trials of molecular memory, namely the building and operating instructions, or genetic mutations; those individuals, that is, which most closely answer to the conditions of their world.

Their evolution, therefore, is a learning process, as Konrad Lorenz has established, a knowledge-gaining process that copies and preserves judgments on patterns of structure and behaviour in the area essential for the species.

What there is to be learned in this world

As will be remembered, we presuppose that there is indeed something to be learned in this world. Indeed, in a world assumed to be chaotic, life could not have arisen, let alone learn something or develop. The fact of evolution is enough to show that the world possesses order. What is extracted from this order is its regularity, the accumulation of contacts and coincidences of its states and events. The question at first is not how the world actually is, but with what reactions we could most easily tackle it, with what essentials we could most simply and surely come to terms with it. Approaching its possible reality is an asymptotic process of optimisation, which will doubtless never end; thus there can never be any compelling certainty.

We may recall how naturally the neck of the femur seems to follow the laws of mechanical loading or the eye of vertebrates the laws of optics (Figs 2 and 3). This, we shall see, is no rare accident. For example, whenever swift movement in water is needed or a complicated eye is developed, then the laws of streamlining and of optics respectively are fixed in the structures. Naturally, the same holds for the inherited programme of behaviour, as we found in paramecium (Fig. 4), nor do any of these organisms know anything about stress-lines, focal points or hydrodynamics. In each case, evolution favours reproduction for bearers of those random changes that move towards the laws governing the conditions of life, which displaces their competitors. This is a structural representation of life-promoting conditions with the "expectation" that these will always remain the same or repeat themselves. It is, therefore, a judgment in advance, a prejudgment established in the molecules of the genetic material, the structural and operating instructions of the organisms. In accordance with the "strategy of genesis"(13) it helps substantially to raise the chance of winning in a search, using chance. An evolution that has to be creative exclusively on the basis of chance simply cannot afford to let the field of search grow without bounds.

The origin of prejudgment

We must now pursue a particularly important characteristic of order in this world: namely, that this order is extremely redundant. This means that its objects and events uniformly repeat themselves many times, indeed unimaginably so. The letter "e" for example occurs in this book roughly 40,000 times; the identical bricks in a town, the individuals of our species 10 to the power 9, still commoner species 10 to the power 12. The grey brain cells of a man repeat themselves 10 to the power 11 times and the red blood cells 10 to the power 15 times. The universe contains 10 to the power 22 stars and 10 to the power 80 quanta. Quite similar numbers apply naturally to events, the formation of these large numbers of identical cells and individuals. Since life arose on our planet 3 x (10 to the power 9) years ago, the sun must have risen and set some 10 to the power 11 times.

One can think of other forms of order which consist only of redundancy and those which contain no redundancy; in which therefore, only one single object is constantly repeated; or nothing is repeated. In both cases the cognitive apparatus as life has developed it would be unable to acquire any knowledge. Its learning mechanism is attuned to separating like from unlike. In most cases, life may depend on the objects as well as events being repeated. Indeed, it repeats them in the continued repetition of generations, reactions, courses of movement, identical words and experiments to be checked. The encounter or the coincidence of an event with a particular life condition will, as a rule, be countlessly repeatable as such or in explorative behaviour.

The probability of coincidences

The coincidences need not be compelling. The learning process has nothing to do with conclusive argument as known since Aristotle and dominant in scientific logic since Frege(14). It is not the necessity but the probability of coincidences that is represented. If, when it is dry, soil organisms crawl deeper into the soil, this does not mean that it must be wet in the depths because it is dry above. The programme is simply modelled on the probability that with dryness in one layer of soil, it will generally be wetter with increasing depth. It is sufficient if the prejudgment of the molecules decides correctly much more often than random trials (Fig. 17).

That the process can produce programmes with extraordinary certainty of success is obvious; and this depends on the fact that, of all possible coincidences between reporting and the life situation, the most constant coincidence is programmed again and again.

Ticks, for example, require the blood of a mammal. Consequently, amongst all natural objects, they have to find mammals, and the most useful instruction will be that which meets this requirement most simply and reliably. Now it possesses an inheritable programme which makes it stop on smelling butyric acid and allows it to drop from the branches and, on touching some object at 37 degrees centigrade, to pierce it. This "definition" of the mammal in the "world view" of the tick cannot be surpassed either in simplicity or in certainty of success. Error is almost excluded.

Connecting by trial and error

The process of such a linkage of coincidences is likewise known in principle. It resembles a development of firm connections in building apparatus by trial and error: the building instructions of all the successes obtained are passed on and those of failures rejected. If, for example, it turns out that the coincidence of news of some resistance at the cell tips of a paramecium with the consequent order for backward waving of the cilia is always successful, then those individuals whose mutations happen to have firmly incorporated this coincidence, are highly favoured by selection and quickly spread their building instructions (Fig. 4). The same holds for encoding of the information "dryness" with the order "creep downwards", or the information "butyric acid" with the order "drop".

The material basis of the programme

From one of the simplest of all organisms, the coli bacterium, we know even the molecular relationship of such a programme in the building instruction itself(15). The section of the genetic material, containing coded instructions for production of the enzyme needed for sugar digestion is closed off in its vicinity, in the absence of lactose (milk sugar). This barring is called a repressor or a repressor molecule which, on combining with a lactose molecule, can no longer act as a barrier (Fig. 7). The entry information which thus consists in the structure of a sugar molecule, is firmly associated with the subsequent information "production of the sugar-degrading enzyme". In the paramecium, the programme is already delegated to the cell plasma by genetic material and in Metazoa, the tick, for example, it is further laid down in a chain of specialised nerve cells which lead from the sense organs to the co-ordination of quite complicated behaviour patterns. We must indeed expect that the whole host of regulatory, motor and reflex programmes, right up to the most complicated hierarchies of instincts (cf. Fig. 37), is programmed through more and more canalised routes, through similar recognition by means of switching patterns.

Experience in retrospect, judgment in advance

The prejudgment of molecules, in any case, depends at all levels on the firm linkage of particular information to decisions which are far more probable than any random attempts at correct judgment. It always arises through trial and error, just like any experience in retrospect, but it contains a judgment for the future, a judgment in advance, that relieves the organism of those essential decisions that chance gave to it only with enormous losses and could offer again only with similar losses. The prejudgment of molecules breaks down the repertoire of chance, reduces the possibilities of senseless trials and hence of nonsense, chaos and ruin. It accepts the repeated conicidences in nature as something apparently essential. It rejects all such hypothetical reality in case of failures but it always accepts as ever more probably true everything that is possibly true, the more constantly and frequently that its prejudgment is confirmed; even though it proceeds with the restraint of utmost economy in experiential gain(16), and incredibly slowly.

THE ECONOMY OF PROBABILITY EXPECTATION

The slowness of learning of molecules and the rigidity of their programme must have become one reason why individual learning is so successful; and with higher differentiation, particularly of the composite sense organs, such as the vertebrate eye and the central nervous system, it has become highly developed. Indispensable as the learning molecules are for evolution, compared with the speed of individual gain in experience it would be catastrophic to have to wait many generations for improvement of a reaction and then to have to drag it along unchanged again for many generations(17). However, no individual learning would have been possible if it had not been able to build on a highly differentiated pattern of molecular experience, which the genetic material extracted from its world long ago. In the realm of behaviour, these are instinctive movements which range from the simple kinesis reaction of which we are already aware (Fig. 4) to the hierarchy of instincts (Fig. 37) yet to be mentioned. Roughly halfway along stands the "unconditioned reflex": for example the eyelid reflex (Fig. 8) which immediately closes the eye when a ray of light strikes the cornea (Fig. 10), or the patella-tendon-reflex, in which the leg immediately extends on sudden tensing of the tendon.

Individual learning

Individual learning depends on an opening of the closed molecular inherited programme(18) and on the formation of a new linkage or combination with it. The centre point of a whole range of forms of individual gains in experience is the conditioned reaction discovered by the Russian physiologist, Pavlov. The dog experiments are classical examples of this work(19). At the sight of food, dogs automatically secrete saliva as the result of a molecularly inherited, unconditioned reflex; but never at the sound of a bell. For bell sounds have never had anything to do with food in the selection area of the dog's evolution.

However, if the food bell is sounded (cf. Fig. 22) regularly enough when they are fed, then saliva will soon be secreted on sounding the bell alone. Two channels of information thus become joined together. Their existence is the prerequisite; their association is something new. The old pattern is just as indispensable as the new, which also becomes indispensable by speeding up the often vital gain in experience. The one remains necessary, and the other becomes a new essential.

In this, stratification of the real world(20) firmly retains its structure. Let us here consider only the connection of the less with the more complex. From the laws of quanta, atoms, molecules, biostructures, the set of laws of each layer extends through all the overlying ones. What laws other than those of quanta should the combination of quanta further contain for the new features of atoms; which for the new biostructures, than those of molecules? This structure continues with the same necessity in the towering edifice of the cognitive process.

The penetration of the building laws

On what, therefore, could new associations between programmes depend if not on the long proven and established programmes in selection in their lower layers? It is thus no longer a miracle, but a shining confirmation of the lawlike links between layers in the world, that in individual gain of experience, in the learning of switching connections, the same learning principles apply as in the learning of molecules.

Indeed, these two levels of learning too are interwoven by intermediate layers, as was shown by Konrad Lorenz(21). Thus, we know of individual reversible changes in inheritance programmes, such as habituation or sensitisation. However, we also know of the formation of associations that are irreversible for the individual, like habituation, trauma and imprinting(22). Imprinting can be interpreted as the completion of an inheritance programme through association with an individual learning experience. It saves incorporating complicated information into the molecular memory, as it relies on the probability that, during some special sensitive phase, the image of an associate, a sex partner or even an enemy would be kept in view, and so fixed irreversibly. In this interlude, too, the operation is characteristically not a matter of compelling consequence, but only of strong probabilities. Obviously, imprinting was discovered(23) in cases of unnatural, spurious, imprinting objects; and it is surprising how varied, even absurd, the objects of imprinting can be (Fig. 9). Evolution could rely on the improbability of meeting them in nature. The transition of associations to probability learning is quite gradual.

Associative learning by the individual now links the constancy of coincidences in the same way into a prejudgment, a prognosis of subsequent happenings, where the confirmation of each single expectation in turn strengthens and fixes the prognosis of subsequent events, but each disappointment or frustration causes their dissolution.

The penetration of learning laws

To look back once more: the mechanism of all creative learning, as we have outlined it so far for the domain of the preconscious, depends on two basic characteristics or peculiarities of this world. One is its considerable redundancy content; the other is its indefinite constancy. That is to say: in most cases we must expect that the same events will recur, but it remains quite open in what circumstances and in what sequence.

With such presuppositions, relating to the whole environment of organisms, a quite distinct algorithm for the learning solution of the problems of life and survival will be successful. An algorithm, namely a method of calculation based on a cyclic repetition of rules(24), which reckons on probable confirmations of indefinite sequences, therefore, is the basic biological principle of heuristics, an empirical method for discovering new knowledge(25). Such a principle of empirical conjecture of possibilities whose presuppositions reside in reflective reason was already outlined by Kant when he says: "Only in empirical science can conjectures (by means of induction and analogy) be permitted, but in such a way that at least the possibility of what I assume must be completely certain"(26).

The biologist recognises in mutation that searching, which we reflectingly experience as the conjecture of a possible solution to a problem, for a success in life. He knows that many confirmations or corroborations are needed before a mutant is established as a definitive experiential gain in the joint learning of the population as a whole; until such a solution is incorporated for all members in the structure of the genetic material, whether it codes for the structuring of the regulators, of the body, of reflexes or instincts. Confirmation must outweigh disappointment.

It is interesting that animals, too, learn creatively from each other in this way. Mere imitation, which so dominates learning in our civilisation alone, we do not include here. Thus, the mimicking "imitators" of poisonous animals must not become too numerous, for then the predators individually learn to become undeceived when it is to the very deception that the imitator owes his by now genetic deceiving likeness. Wickler has made this very clear(27). Again, plants that have flowers of one sex and are pollinated by insects can cease to produce nectar only if the female flowers become extremely like, but not more numerous than, the male ones. Many details of such "deception" flowers have been clarified by Stefan Vogel(28). Thus in Begonia species, where 70% of the male flowers, by their pollen supplies, can retain foraging bumble bees with adequate food. The same, in the opposite sense of training, is known from Kuyten's work(29). The caterpillar of an emperor moth from Assam, before pupation, spins itself into a rolled decaying leaf. To do this, it bites through the leaf stalk and spins together the leaf. So that birds do not become trained to this "food roll", however, several leaves are always bitten through and spun together and some two-thirds of them will remain empty.

The penetration of the algorithm

As to the individual learning of predators, such as bumble bees and birds, on the teaching side it is still the learning resulting from the teachable genetic material. The principles of disappointment and confirmation therefore continue unchanged. We already know this about individual learning from conditioned reflexes, which likewise correspond to heuristic learning. In fact, Pavlov's dogs had first to discover the connection between bells and food. From the work of Grant and Schipper(30) on the eyelid closure reflex we even know that the relative frequency of reinforcements in the conditioned reaction shows in the size of the learning success. Before the conditioned stimulus, which was a stream of air onto the cornea, a flash of light was used as an unconditioned stimulus. The maximum of positively conditioned reactions (Fig. 10) reaches the relative frequency of the reinforcements. Even unlearning takes place more quickly the more constant the confirmations in the learning process, as if the disappearance of the most unequivocal connection were recognised most clearly by the nerve association produced. Klaus Foppa has given the most comprehensive treatment of such findings(31).

However, no higher organisms simply wait for the onset of definite stimulus situations, but set about searching for them, according to the guidance of their own inner needs and moods at the time; just as when we want to release the stimulus of sneezing, or to satisfy that of hunger, we blink into the sun or set off to look for an inn sign. Since the work of Craig we speak of desiderative or appetitive behaviour(32). This gives rise to a new promoter of learning. In fact, Pavlov's conditioned reactions do not turn out to be conditioned reflexes as the great physiologist intended, but conditioned appetences. If the dog in question were freed from the chain, then, on hearing the bell, he would jump up barking and wagging his tail and, as Bernhard Hassenstein says, thereby shows, "as a result of learning, his typical appetitive behaviour, namely social begging for food".

However, the conjecture of possible connections goes further still. To quote Hassenstein again: "An originally neutral behaviour or element of behaviour is placed at the service of appetite behaviour." Karl von Frisch kept a ". . . little parrot in his room. He allowed the bird to fly about freely for some time but only if he had observed that the parrot had just deposited its droppings in the cage; in this way the room would remain free from 'visiting cards'. The bird now soon learned, when he wanted the reward, to produce small amounts of droppings even without the inner necessity. His efforts in this respect had unusually comical effects. The straining became for him an act which paid off, and from time to time he began to ask in this original way whenever he saw a tasty morsel or had some other lively wish, even outside the cage." Or: "A further example is seen in the agitation that a small ape in a zoo created, who was being pushed aside by stronger animals; he began to jump up and down on the spot. Visitors to the zoo had their attention drawn to him by this and they now threw food to him over the other apes. Thereby the drive to acquire food was linked in the animal with 'successful' behaviour. The greater the hunger, the more frequently he now jumped up and down on the spot"(33).

We shall indeed observe that our own actions, even in the repetition of scientific experiments, are guided by appetite behaviour; and that the conjecture of possibilities, using neutral substitute relationships, leads to superstition. If we touch wood in good time, have we not usually averted the worst and accordingly repeat this strange behaviour? Paul Watzlawick provides amazing examples(34) of this part of our "reality".

The algorithm of conjecturing possibilities is therefore extremely old, in the learning of molecules it is as old as the genetic code; 3000 million years. However, individual learning of the switching connections must also be as old as the complex nervous systems, probably 500 million years(35). That is 500 times as old as the genus Homo and 5000 times as old as Homo sapiens neaderthalensis, the oldest form of our species.

The ratiomorphic apparatus

Within this enormous time span of individual learning of linkages, a linkage network has developed in the brain, which although still a long way from any conscious deliberation, now achieves quasi-rational results on this intermediate level. Along with Egon Brunswik, we can speak of a ratiomorphic apparatus(36). We know of its astonishing accomplishments in examples from the whole range of vertebrates; this includes even sharks, the ancestors of which are separated from ours by almost 500 million years. They achieve quite appropriate, functional results in complicated stereometric calculations, which, as in man, enable them to estimate their own movement relative to that of an object, or the most varied retinal images that differ appreciably according to distance and perspective, so that they can decide reliably and again in advance whether it is the same partner or the enemy. The calculations are indeed so complex that even the most elaborate computer program has not yet been able to duplicate them(37); this is enough to show that our conscious, rational reflection has indeed overtopped ratiomorphic reflection but certainly not replaced it.

Consciousness and conscious rationality, those significant achievements of man's evolution, have their own laws and form, a regulator or control organ for unconscious ratiomorphic performance, but they also form the youngest organ by far of the knowledge-gaining systems in living creatures. As already mentioned, it must have developed when our forefathers became erect and acquired tools, together with the development of language and the handing on of individual learning. That was about some five million years ago(38), one hundredth part of the time during which the ratiomorphic apparatus had already existed.

The overestimate of the rational

People tend greatly to overestimate the rational portion in our achievements. This is quite natural, for we are aware only of the conscious and it is conscious reason that is specifically human and places us above the animals; oddly enough, civilisation's technically organised contrivances impress us more than our having survived in spite of all its flowerings. Leaving aside this self-admiration and all that we think we owe to ourselves alone, the ratiomorphic achievements remain. They show that most associations are still linked in the unconscious, that their drive remains detached from consciousness, that all that is creative(39), happens beyond the conscious, as Arthur Koestler shows; and all we possess in memory content, complex solution(40), combinatory theory, configurational experience(41), must, as Konrad Lorenz observes, first be brought out from the unconscious, indeed be dispatched laboriously into consciousness, to sink back only too readily into the unconscious. Consciousness is a thin layer over the 1000-million-year deep substratum of its own unconscious presuppositions(42). To this extent Freud and Jung are completely right.

The simplest ratiomorphic operations

About all this, more later in the appropriate place for each control mechanism. Here we shall first of all pursue what this ratiomorphic apparatus contains in the way of simplest operations. The foremost presupposition to be observed is again the assessment of probability.

Again we have the probability of the world's real existence, which, like an assumption or hypothesis, occurs right up to the conscious level. Once more, all those coincidences count as probably real about which predictions or advance judgments are confirmed in the event; as we shall see, the degree of probability increases as a power of the measure in which expected confirmations recur without contradiction. On the contrary, as in the learning of molecules, whenever a hypothetical forecast fails, the expected degree of probability decreases. The amount of what is predictable, the abundance of characters, like the repeatability of prognoses about the world, is so unimaginably large that for this reason alone the reality of the whole becomes extremely probable. If I open my hands, this book will fall down, or if I glance up, the space around me will still be present, or if I close my eyes, the sensation of brightness will be changed into a thought; these are all tiny building bricks for that gigantic edifice which contains our expectation of a real world. Under what other assumptions could the ratiomorphic apparatus have asserted itself and have guided its bearer successfully through the dangers of his world?

Just as life in general, it is a hypothetical realist. Donald Campbell first coined the term "hypothetical realism"; and he has applied it to the whole epistemological attitude that we are adopting here, outlining its biological history(43). Actually, it takes consciousness to doubt that the world is real, namely when it disregards its ratiomorphic background. The dilemma of consciousness is that it has no direct knowledge of its background. What is "the backside of the mirror"(44), as Lorenz shows, was grasped only after long research.

Assessment of chance versus necessity

The most fundamental assessment that the ratiomorphic apparatus continuously performs, however, is not so much concerned with separating the real from the non-real. These are rather alternatives from within the layers of consciousness. In the first place the ratiomorphic apparatus differentiates between the fortuitous and necessity. The hypothetical reality of these alternatives is then a simple consequence of the fact that the calculation is forced to assume a world of necessities.

The mechanism, or better the algorithm(45), at the basis of this calculation, is again the one we know from the learning of molecules, from conditioned reactions and even from imprinting. It again depends on the countercalculation of probabilities. We are now so close to consciousness that the method of calculation can enter our sphere of conscious observation and be formulated rationally in the form of language and mathematics. We shall compare the results of such formulations with our pre-rational perceptual judgments so as to verify that the most rational mode of reckoning of apparent truth corresponds exactly to the method of the ratiomorphic mechanism.

In its very statement, the ratiomorphic mode of reckoning behaves as if all states and events could be neatly sorted into chance and necessity. This in itself is interesting, since we need not know what chance is, nor whether it may exist within our awareness and hence outside microphysical phenomena altogether. Viewed subjectively, from the position of the individual who must decide, this sorting is very rational because it includes the presumption that one will have foreseen the re-occurrence of some states and events but not of others; the problem of orientation in this world depends on whether one can possess foresight, or lacks it. As yet there seems no third option(46).

The hypothesis of apparent truth

If we analyse the content of this expectation, we find three levels of supposition: the assumption that certain events may probably be observed again, therefore the assumption that forms of order in the world recur, and hence the assumption of an apparently real world. For the hypothesis of apparent truth contains the expectation that under suitable conditions some experience may be predicted as probable, and therefore confirmed by occurring again. Let us examine next what is to be processed here under the heading of "probable". Here we find that neither the date nor all the preconditions of an event can be predicted, but its degree of possibility can. This corresponds to the hypothesis that one could have some foresight of the scope for manoeuvre granted by the world to its chance components. That is all the more remarkable since there is no inkling at all of whence this advance presumption could stem; because, if we reflect, probability for us is a measure of the degree of possibility of states or events still unrealised. The probability of winning corresponds to the reciprocal of the domain of chance, namely the number of possible choices that chance can make. How could one know in advance the extent of an unknown domain? This amount of ignorance, about which the hypothesis of probability presumes to have foresight, may also be described as the ignorance of the domain of chance. This will be briefly illustrated, for it corresponds to the core of the three axioms(47) underlying current mathematical probability theory. For the case where all the chance events that can occur within the bounds of a given condition are equi-probable, the probability of one of them must correspond to the reciprocal of the domain contained in the condition. The domain under the known conditions for coin tossing, throwing a dice or dealing a pack of bridge cards is 2, 6 and 52; and accordingly, the probability that "heads" turns up or the one on the dice is thrown or the Jack of Hearts is drawn is 1/2, 1/6 and 1/52 respectively.

Even by now, we have used several different concepts of probability. In our numerical example we are dealing with a posteriori probability, a frequency interpretation after the event. If we take dice, for example, only an endless number of throws will allow us to determine accurately the probable frequencies of the faces. In the creative learning process, it is a question of a priori probability, hence a judgment in advance; and, moreover, one of an extremely subjective form. For, as we know, this advance judgment must be able to start from every kind of ignorance.

A subjective probability expression was developed in the thirties by Frank Ramsey and by Bruno de Finetti(48). It "represents a more precise and idealised and rational model for the pre-scientific, intuitive concept of probability that we use in everyday language" and is intended, Franz von Kutschera goes on, "to develop a general framework for linking rational assumptions of belief"(49). The forms it takes, as well as the objective and logical probability concept to be compared with it, will be discussed later. Here it will suffice to observe how the subjective probability concept as it applies to the biological learning process agrees with and differs from the normal concept used in the literature.

The agreement resides in the unconditional but nevertheless guiding function in the knowledge-gaining process. Here as there we observe that "the intuitive evaluation of events by probabilities serves, in many practical cases, as the basis for our actions; where we have no certainty as to whether event E will occur or not but the success of our actions depends upon it, we will be guided by the probability that we assign to E". This is primarily the sort of case where, for example, our expectation is based on "that it will not rain": if we are meteorologists, we follow the weather forecast; otherwise the farmers' rule of thumb, because carrying an umbrella is either tedious or we do not have one handy. However basic the agreement, so is the difference. The formal concept of subjective probability, as von Kutschera sums it up, "neither tries to reflect the actual belief of someone or other, nor the principles on which these beliefs are validated. What is actually believed and how, is a purely empirical question"(50); and this empirical question is ours. We are not dealing with logistic but rather with heuristics, not with deduction or compelling derivations but with induction or possible construction; the cycle of the knowledge-gaining process, as described by Erhard Oeser(51) concerns not formal scientific explanation of proofs but the scientific, experience-based explanation of expectations, a "heuristic probability".

Although we shall find that the forms of heuristic probability run parallel to those of the logistic kind, nevertheless the former are only the mirror-image antagonists of the latter, in the cyclic process of expectation and experience. Whence, we must further ask ourselves, stems our expectation that we have foresight of the unknown, the uncertain play of the possible?

An a priori feature of reason

Actually, no judgment on subjects beyond our experience can be rationally based. Yet we have seen that, without judgments in advance, no experience can be gained at all. The expectation of possibility or impossibility, chance or necessity, is a prerequisite for all acquisition of experience and cannot be based on the individual experience of any one of us. This already confirms Kant, who included expectations under the category of modality in the a priori of pure reason(52). They have remained unamenable to question by pure reason. To judge a priori means, as it did with Kant, to judge "in advance", to judge "without the object being present", a pure self-activity determined only by itself. Without this original spontaneity we should not know anything a priori(53). Still, the origin of the a priori remains a puzzle, although we shall see later that Kant already surmised the solution of regarding it as a "kind of preformation system of pure reason"(54). For us, the solution is obvious; it was already recognised by Konrad Lorenz and afterwards by Donald Campbell(55). The necessity for conjecture of the possible is certainly an a priori for the pure reason of the individual, but it is the learning result of the ratiomorphic apparatus, which depends on the switching instructors and beyond that on the molecular ones; it is an a posteriori of our species. Now we see that however scanty the knowledge of something may be, every decision must be carried by an expectation which, consciously or not, starts from a completely subjective probability. What justifies these prior assumptions is in our terminology the experience that the size of this hypothetical probability is not very important. It is easily corrected by each sequential experience. What is decisive is that we count on a probability at all, that we meet each experience of a sequence with an expectation, a prejudgment, a hypothesis. The economy of prejudgment depends precisely on this.

Two examples: If we draw the Jack of Hearts from a pack of well-shuffled bridge cards, it is not a matter for astonishment. It is part of the domain. However, if after reshuffling the cards, we draw the Jack of Hearts a second time, that begins to be remarkable. How does our expectation change if again we still draw the Jack of Hearts at the tenth or the hundredth attempt? We should have to abandon the hypothesis that we are handling an ordinary pack of bridge cards and perhaps assume that the pack contains only Jacks of Hearts. Our hypothesis of its domain will change from 52 to 1. Or: If we take a bag of A and B letters intended for supplementing our letter case and need to bring out an "A", this will not change our expectation that every second time we shall get a "B". However, if in our next dip an "L", a "Y", a "D" and then again an "L" turn up, we shall begin to think, mistakenly, that we have got an entire alphabet. Our hypothesis of its domain will now change from 2 to 26. All knowledge-gaining operates on hypotheses of expectation, as Erhard Oeser has demonstrated, right up to the level of scientific enquiry. The more precisely the expectation is formulated, the more conclusive will be the answer. He who does not ask cannot experience anything, as Jakob von Uexküll has pointed out. An undefined hypothesis, says Karl Popper, or a vague idea cannot be accurately tested, neither exactly confirmed nor refuted(56).

Only in the "unordered" region of unconsidered everyday consciousness do we imagine we can find our way about without any defined hypothesis. This merely depends on our ratiomorphic apparatus not only continuing to relieve us of defining hypotheses, but also to process them along with the successes as well as the failures of its expectation, in order to keep them continually reformulated, up to date. Egon Brunswik foresaw this, while Konrad Lorenz established it biologically and Gerhard Vollmer epistemologically(57).

A rational algorithm

If we now translate the pre-conscious process of knowledge gaining into the rational mode of expression, then we have the following situation. The processing method is cyclic and leads in a finite number of cycles to the solution or the decision. It has the form of an algorithm.

Nothing is in itself certain. Rather we must constantly reckon with the possibility that events will prove to be either unpredictable or predictable, or as we say, fortuitous or essential. Moreover, certainty is never absolute, we can only more or less approximate to it; we say, it will approach 0 or 1. In case of complete ignorance, all possibilities must be equally probable. This is the state of the highest uncertainty. Higher degrees of certainty, however, will emerge from cyclic calculation of sequential experiences. This calculation was discovered by the English Presbyterian minister Thomas Bayes, published posthumously, but its significance was grasped only more recently(58). We ourselves have developed it, independently of Bayes, from the behaviour of biological systems(59). Now, from experience, we expect that occasionally it fails to satisfy one of the alternative hypotheses, reducing their probability exponentially, somewhat as in physiology today; the ratio of sensitivity to stimulus is described as an exponential function, by the laws of Weber and Fechner(60).

In short, we expect that, with alternative expectations, the number of failures will be as the negative power of the hypothesis of the domain. That this can be accomplished rationally, namely that it does correspond to our personal expectations, must now be shown.

If we engage in a game of chance involving tossing coins, then we can reckon with a chance domain of 2, namely heads or tails, and a chance of winning of 1/2; even if our partner always throws the coin and, let us say, it falls heads. If the first time it is heads, then that will be regarded as a normal consequence in the game. If it does so a second time and then a third, we may still present this to the opponent as luck, even as extraordinary luck. How often, we may ask, can heads be thrown in an uninterrupted series before we have to revise our hypothesis that we are dealing largely with chance, and before we are driven to the alternative assumption that it is no longer a normal matter or chance happening but a question of necessity, purpose, plan or intention? Calculation will tell us. In fact, the probability that heads will turn up either twice or three times is 1/2.1/2 and 1/2.1/2.1/2, that is $(1/2)^2$ and $(1/2)^3$; which means 1/4 and 1/8 and signifies that such events will occur by chance approximately every fourth and eighth game; which, however, is still a possibility. If now heads still turned up on the tenth or even the hundredth throw, the probability that my opponent is not cheating is only $(1/2)^{10}$ and $(1/2)^{100}$, which is equal to 1/1024 and $(1/1.3) \times 10^{-30}$ (compared with this, the universe has existed only 10^{17} seconds). A quintillionth part equals an impossible chance. We shall have abandoned the chance hypothesis long ago. How absurd the world would be if this kind of chance existed was shown by Tom Stoppard in the first act of "Rosencrantz and Guildenstern are Dead"(61), where in spite of all their antics neither of the two succeeded in turning up heads even once. Our corresponding, imaginary, coin-tossing experiment in the lecture theatre, on the other hand, proved that after the fourth, fifth and sixth times of heads being thrown, 80, 90 and 92% respectively of the students no longer believed in chance and loudly interrupted the unexpected trick(62) (Fig. 11); although there were always some who held on unswervingly to the hypothesis of chance. More about this later.

The same holds for the expectation of necessity or of intention. If we are convinced from the beginning that our partner is cheating, then the first throw still tells us nothing. If, however, in spite of that, my prognosis that heads will always turn up is repeatedly disappointed, then I must likewise abandon this suspicion.

A degree of truth or certainty

What changes in our thinking in such "experiments" is the measure of truth or certainty that we attach to our judgment, hypothesis or expectation. Our attitude shifts from uncertainty or perplexity to a state in which we are prepared to assert, to defend and finally even to risk increasing wagers. This again is mirror-image behaviour between our heuristic-inductive and the already mentioned formal-deductive treatment of probability(63), of which more later.

The shift in our attitude may indeed be measured. However, we do not want to speak of a degree of truth. For truth is an overloaded concept; it seems as if it could exist without the observer; and as truth value and logical truth it figures in logistic(64), where it again appears as if this world could be divided into true and false; actually, however, logic describes only the faultless transfer of truth, a truth of which we think that no one truly possesses it. Here, on the other hand, we have to do with the discovery of relative truth by erring subjects and therefore we had better speak of a degree of certainty (G).

What we assess is a quotient from the probability (W) of possible chance (W_Z) and possible necessity (W_N), numerator and denominator being raised to the power of the number of failed expectations (e). The degree of certainty of chance expectation (G_Z) is then $G_Z = W_Z^{e'}/W_N^{e}$; the degree of certainty of expected intention, determination or necessity is its reciprocal. If, as mentioned at the start and as is usual in logistic and information theory, we wish to have values of between 0 and 1, then one writes $G_Z = W_Z^{e'}/(W_Z^{e'} + W_N^{e})$ and $G_N = W_N^{e}/(W_N^{e} + W_Z^{e'})$. In investigating a sequence of events we may indifferently start from the expectation of chance (G_Z) or from that of necessity (G_N). For in the case of complete uncertainty, both are equally uncertain, but must lead to the same result in each instance. Finally, it should be remembered that all probabilities (W) change simultaneously with experience and correspond to the reciprocal of the expected domain; and that the sum of the failed expectations of chance (e′) and necessity (e) corresponds to the number of prognoses tested or repeated observations respectively in the sequence of events. Only in the case of very small domains is e′, the number of failed expectations, to be related to the value of e′ — W (65). In the result, 0.5 indicates maximal uncertainty, an approximation to 0 or 1 high certainty of the predominance of chance or necessity.

Experience and hypothesis

An unequivocal determination of expectations that have been reckoned as failures requires an unequivocal determination of what the hypothesis contains. Since nothing can enter the expectation of a chain of unknown events except the experience gained with it up till then, we may define(66), in the case of expectation of necessity (G_N), the shortest sequence so far confirmed as a measure of the hypothetical expectation of the sequential event, for example, "1-2-2" from the series "1-2-2-1-2-2-1". This hypothesis will change with the development of experience and finally be continually confirmed or be refuted. In the case of chance expectations (G_Z), the alternative of the expectation of necessity must again be forecast. This means that the sequential event is to be defined from the non-periodic or non-cyclical continuation of the series(67).

Further, we may keep in mind that, in the sequence of confirmation or disappointment of our expectations, it does not matter whether the contents of our expectation appear simultaneously or successively, side by side or one after the other. It will be admitted that the simultaneous turning up of ten heads can be no more probable than heads turning up ten times successively: in the exercise of our observation, however, we understand the simultaneous nature of the coinciding events as an abundance of characteristics, while the successive coincidence counts as mere repetition. However, both of these, both simultaneous and successive coincidence, multiply each other, if it is a question of separating chance and necessity in this world. In connection with this hypothesis of comparison we shall later show how most objects of our attention are extraordinarily rich in features. Besides, in ordinary life as in research, we try so to arrange matters as to be able to observe the objects we are interested in as often as we please.

With the insight that confirmed expectations of simultaneous and successive coincidences together determine the degree of certainty concerning the rule of regularity, we have progressed another step. This will prove to be the basis for all comparison and inference. We have thus not only introduced the determination of subjective degrees of truth or of certainty, but also laid the foundation for a general theory of comparison which, surprisingly, science does not yet possess. It was structural research in biology, namely morphology, that first attempted to realise this in the so-called homology theorem, for which Adolf Remane(68) has given special criteria. We shall recognise that, within the framework of the "hypothesis of the comparable", we have before us the first solution of the homology problem.

In this mirroring of simultaneous and successive coincidences, there resides, as we shall see, that no less remarkable differentiation between structural and causal laws (relating to the former and latter respectively). This has to do with the differing competencies of our "innate teachers" in the preconscious processing of complex data. These we shall meet again.

At last, we can confirm that subordinate significance of the starting hypothesis with which we had to begin: what scope can be assigned, in particular, to the play of chance, to the domain in as yet unknown phenomena? Of course, after drawing the Jack of Hearts five times from the 52 bridge cards, we are more convinced that some intention rules than after turning up heads five times when tossing a coin, for $(1/2)^5$ still corresponds to a probability of 0.03125, while $(1/52)^5 = 2.6 \times 10^{-9}$, which is as good as impossible. In nature, we nearly always have to deal with phenomena that allow for at least ten characters each on ten repeated observations. With starting hypotheses of that kind, the chance coincidence is already practically nil. If, very generously, we allow a probability of 1/2 for the chance appearance of the individual character, then the chance expectation of what is observed is still only $(1/2)^{10.10}$, which equals $1/1.3 \times 10^{-30}$, practically an impossibility. With such impossibilities it does not matter how impossible the impossibilities might become. Therefore it suffices that this basic hypothesis of the ratiomorphic calculation operates with probabilities at all in order to develop the degree of apparent truth always by learning from tested predictions.

Heuristics and probability

If we could follow the concepts of ordinary speech, then this algorithm of knowledge gaining would have to be called a probability logic. However, scientific logic, since Gottlob Frege's "Begriffsschrift" of 1879, was no longer viewed by Russell and Whitehead as a "theory of thought"(69) but has become a "theory of the forms of true statement". It is no longer concerned with the acquisition of truth, but investigates conditions for the suitable transmission of assumed truth. It has abandoned the field of heuristics, namely the study of methods for discovering new knowledge, or is now trying, as Carnap and Popper have done, to establish laws of truth discovery via its very precise laws on truth transmission(70). It has abandoned inductive logic, inference from the special to the general, and has limited itself to deductive logic, drawing conclusions from the general to the specific, where alone binding conclusions are possible. On the contrary, with heuristics, we are concerned with induction and concept formation and can profit only indirectly from deductive logic.

Heuristics, which primarily is the creation or discovery of new certainties by means of an algorithm of probability, is the method used by living organisms right up to the ratiomorphic instructions of human reason. We may recall that, up to the learning of the instincts through genetic material and to the individual linking of coinciding or conditioned events, the extent, accumulation or probability of coincidences guides the formation of associations. "On the other hand", to quote Klaus Foppa, "the complexity of the reflected event appears to have only slight influence on the rapid establishment of stable conditioned reactions, since in the normal surroundings of living organisms, complex ways of behaviour can be conditioned quickly and permanently"(71).

In this general connection, one must keep in mind the circumstance that it is certainly the redundancy of phenomena of this world on which the algorithm of biological learning depends, but neither recurrence nor absence of confirmation can be expected with necessity. The succession of confirmations or disappointments in all objects of creative learning in nature can only be the size of a probability. "The fact", to follow Klaus Foppa's important insight, "that is learned under relatively fixed conditions requires no explanation, but that progress in learning is possible in spite of variable conditions."

In other words, reinforcement always involves two classes of reaction: the event either occurs or not. Investigations of this kind into probable reinforcement go back to Brunswik and Humphreys (at the end of the thirties) and to Foppa, who ten years later was able to assemble abundant material on the "probability model of learning"(72).

A typical study is that by Grant, Hake and Hornseth, who set test subjects the task of guessing the sequence of illumination of a lamp every five seconds following the flash of a starting light(73). The reactions of persons taking part quickly settled down to the asymptotic course with which we are already familiar, constant light or constant dark (Fig. 12). At 25 and 75% of illumination frequency, they only slowly adapted to this. With the same frequency of light and darkness, they remained at 50% positive reactions. That is, they were least certain of the probability of sequential events.

We ourselves have continued with experiments, in which the persons taking part have to decide from a chain of events, whether and with what certainty they expect that a chance series or a programmed one were involved and, hence, whether it involves intention, determination or necessity. The procedure was to extend the chain of events for the participants, who had to record their prognosis on a form before each extension(74). This method of continuing a series is related to the "non-verbal intelligence test", the analytical test of Meili, as well as to Amthauer's structure test. Simon and Kotovsky have investigated optimal solutions and Krause the optimum amongst the strategies available for problem solution(75).

The forms of heuristics

Some interesting results of our experiment are, first, that there was no hesitation to write down forecasts as soon as the first event had occurred; secondly, that the mean value in programme discovering asympototically approached the correct result, with decreasing standard deviation and, thirdly, that its course followed essentially that of a completely rational solution process, while deviating significantly in its growth of certainty. This is particularly the case with regard to the task of recognising a sequence of events as due to change (Fig. 13 right).

This empiricism of heuristics thus reveals a behaviour related symmetrically to that of forms of probability in logic, already defined by the theory of science. In this process of prescientific gaining of experience like the "process of empirical science", it "always has a cyclical character", as Erhard Oeser puts it; these forms of expectation are opposites, like induction and deduction, heuristics and proof theory, like the hypothesis of the superior principle for the prognosis of its subordinate cases(76). This symmetry goes so far that for heuristics we can use the terms already introduced.

On the way to certainty, the participant in the experiment begins, in complete ignorance of the experience to be expected, with a purely subjective probability, made up of some sort of assumptions close to total uncertainty. The subjective heuristic probability differs from that in logic in only one, but nevertheless fundamental point. We have already met its formulation by Finetti, Ramsey and Savage(77). The heuristics of subjective probability, on the contrary, is concerned with every person's belief and with "the principles valid for this actual belief"(78). A wonderful kaleidoscope of mistakes is revealed if one asks the participants about the reasons for their first decision(79).

With many repeated confirmations of an expectation, whether it be the uninterrupted extension of the "same" programme or the chance distribution of the results in the coin tossing experiment, certainty increases and the prognosis becomes more objective. The probability that something other than expected would emerge disappears, subjective arguments recede, and it looks as if we had only the characteristics of the object, independently of the person judging. This is objective probability. It "should be a statement about the nature" of a thing and "not a matter of expectation by a subject". So much for agreement with the logic of proof theory. This goes back to Bernoulli, to frequency interpretation, the theory of Charles Peirce, Popper and Hacking(80). The difference depends, of course, on the fact that in heuristics, statements must always begin with subjects, even if they are hardly about them.

Objectivity is achieved by frequently observing the processes of solution. The algorithm of the most successful problem solution can be defined. Since the work of Keynes, Jeffreys and Carnap's many studies in logic, this corresponds to logical probability; the evaluation of probability, "which would be put forward by a completely rational person who is free from prejudice and who makes no arbitrary, unfounded assumptions"(81). This best prognosis from experience we have already defined as an expectation that counts on the recurrence of the shortest and least redundant sequence from previous experience. The results of Simon and Kotovsky, as well as those of Krause, on the psychology of thought, point in the same direction(82). So much for the expectation of necessity. The expectation of chance contains the converse. Agreement with logical probability in heuristics is again considerable. Particularly if we see its development as a process; and if by "rational" we mean successful. Of course, the difference is that in heuristics there can be no entirely rational solution. In it, every kind of reason is the product of creative learning, which can never be complete.

In any case, in heuristics, it is always a question of judgments on probabilities. These, however, change with experience from subjective and irrational to objective and rational structures of the prognosis (Fig. 16).

Ratiomorphic versus rational solution finding

We use the terms ratiomorphic and rational in the sense of unreflective and reflective behaviour. Thus we set aside the to and fro of valuation that the term rational has gone through in the history of our culture, led on by our growing admiration for our own rationality. By reasonable and unreasonable, we mean the possible forms of this reason according to its success in the process of creative learning. We can thus compare the relative reasonableness of ratiomorphic and rational behaviour.

As the process of finding solutions with test subjects showed (Fig. 13), it clearly deviates from the hitherto most reasonable solution pathway. For the degree of certainty increases more quickly with the discovery of some regularity, while with that of a chain of chance it grows much more slowly than is reasonable(83). Recognition of a chance sequence ratiomorphically may be simply the result of fatigue. This hankering for the lawlike is indeed a mistake well known to the psychology of thought(84). When asked about their method of finding solutions most of the test subjects stated that, if it facilitated a solution at all, certainty was reached on the whole more slowly than is possible rationally. Once more, this confirms our assumption that the by now familiar a priori of ratiomorphic probability calculation is a wise guide.

Purely rationally, the finding of solutions is hindered by two blind alleys, which open up if we attempt to escape from the ratiomorphic intuitive teacher. For chance can be mistaken for necessity and necessity for chance, quite rationally, and incorrigibly. Even the trick with the coin tossing experiment (as will be remembered from Fig. 11) has shown that some test subjects either from the very beginning believed in design; or, in spite of repetitions, would not abandon the conviction that it was a matter of chance. These probability syndromes are rationalised as follows.

In the chance syndrome, it is asserted that every throw of a coin contains alternatives of equal value and, therefore, every possible sequence of events would have the same probability; throwing heads ten times, no less than each of the others in ten throws; so 1024 different series are possible. This is indeed perfectly correct. The probability for chance will fall off exponentially only at the moment when the prognoses of subsequent events are confirmed. Miracles feed on the chance syndrome, of taking design or necessity for chance; from the suburban magician right up to demagogy.

In the necessity syndrome, it is again asserted that even with the longest chain of events that do not recur in sequence, it could be a question of a programme that is arbitrarily long and thus incomprehensible. That too is correct. It could involve the contents of world literature, coded in yes-or-no decisions. Here again, ratiomorphic teaching is overlooked, which would advise us to rely only on the predictable. To be sure, necessity has time and again been taken for chance; hieroglyphics, for example, for ornaments. However, acquisition of knowledge can prosper only when we discover the predictable from the, as yet, unpredictable. Again, miracles feed on the necessity syndrome, of taking chance for design or necessity; from superstition right up to demagogy.

What can we learn from this? We recognise that solution finding by reason and by the ratiomorphic apparatus must argue differently since they follow different courses of solution and make different mistakes. Karl Popper's expectation that what is logically right must also be psychologically right(85) cannot apply here. Logic is a product of conscious reflection, whereas the psyche is guided both rationally and ratiomorphically. The rational and the ratiomorphic apparatus are, of course, not independent of each other. However, they have been selected under such differing control conditions that they simply cannot operate in the same way. Certainly reason can imply an overcoming, an adaptation of reason to an extended range of objects; those, in fact, elucidated by reflection, for which ratiomorphic reason was not created. However, in the enlarged world view of mankind, each by itself makes more mistakes than they would make when working together.

Unreflective common sense is not enough for mastering the problems of our consciousness, neither is reflective intellect if it were forsaken by its innate teachers. We consider this to be one main cause of the dilemma of human reason. Perhaps it is simply the problem of mankind.

Foresight maintains life

Without knowing the limits of what can be foreseen, life would be impossible. All gain of knowledge from the simplest structures right up to the most complicated behaviour patterns, contains limits to what has so far been recognised as predictable. To a man who continually confuses chance and necessity, our civilisation usually takes care to guarantee survival by means of the madhouse. Left to himself he would succumb. A paramecium that insisted on penetrating the next obstacle, or a tick that took to formaldehyde instead of butyric acid, would be lost, just like a mammal that forgot the imprinted picture of the enemy. In all these cases, what is necessary and regular is abstracted from what is unforeseen and seemingly fortuitous.

Obviously this is continued in individual learning. Bad experience with individuals of a predator species, good experience with some life circumstance or other must be extrapolated to the whole predator species and to the type of life circumstance in general. This is of vital importance, for animals as well as for man.

When this is recognised, the function of the ratiomorphic algorithm may also be considered established by the "apparently true". The economy of this judgment-in-advance rests on the chances of growing larger. This is evidently not a matter of coin experiments nor of riddles involving artificial chains of events. However, it similarly involves abstracting what is necessary from the fortuitous; and within what is necessary, from the order of this world the recurring universal, lawlikeness, concepts and diagnoses(86). Knowledge of the law contains the life-maintaining gain of knowledge; its repetition allows us to reach certainty.

Inductive inference

With this insight, we have reached the venerable problem of induction. It involves inferring from the special to the general. Since it concerns conclusions, and since binding conclusions are surely superior to the rest, its investigation becomes a matter of logic, and that is where errors begin. The matter deserves our attention. For, on the one hand, there is the view that all empirical knowledge, or indeed all knowledge acquisition, depends simply on inductive inferences and is based on its justification. On the other hand, logic finds that inductive inferences can never be binding. Notable examples from logic are the swans and the ravens. We shall return to this presently.

First, let us recall that the problematic nature of induction was discovered by David Hume round the middle of the eighteenth century(87). Kant was impressed by it. The full range and force of Hume's objection, according to Wolfgang Stegmüller, has become clear only today. Moreover, "every new scientific discovery and every fresh philosophical treatment of induction, seems increasingly to confirm the assertion of the philosopher, C. D. Broad: "induction is the triumph of the natural sciences and the disgrace of philosophy"; and, Stegmüller continues, "since the second half, at least, of this assertion is beyond doubt" one may well ask "does the first half of Broad's thesis also hold?"(88).

Certainly, even today, the greatest philosophers argue about the problem of induction and how it might be solved or even whether it can be solved at all. What has natural science contributed? In effect, nothing! We can take two strange examples from biology; although biology alone already contains fifty million conclusions for the founding of the natural classification of organisms(89). What, then, is the problem of induction?

First of all, the example of the swans that Karl Popper uses: can the number of however many white swans, which we have seen (in the northern hemisphere), ever justify the conclusion that all swans are white(90)? Experience teaches that this is not so. On the contrary, a black-necked swan and even a completely black swan(91) have been discovered in the southern hemisphere (Fig. 14). Now we can deal more accurately with Hume's objections as given by Stegmüller: "What sort of arguments lead from the observed to the non-observed?" for "the contents of the statement in which we communicate our alleged knowledge about the non-observed is not included in the contents of our observational knowledge".

As we recall, the conclusions of logic, since Frege, are truth-preserving; for the whole subject has retreated to the investigation of truth transmission. Therefore, these are not truth-extending conclusions. Thus, Hume's induction problem becomes "are there truth-preserving inferences that extend truth?" His "answer to this question is: no, nothing of the kind!"(92).

So the matter still stands in spite of a considerable literature, in spite of the important studies of Carnap, and of Popper, which have all tried to find a way out of the dilemma; for, in the last two centuries, science has advanced all the more quickly.

The cycle of gain in knowledge

In contrast, we have learnt that all biological knowledge gaining depends on a cyclic process, right up to the exercise of common sense. Parts of this cycle we know as expectation and experience. The tendency to expectation is innate, experience is acquired and, after proving itself, it is inherited genetically and culturally. The cyclic process of this principle of problem solving is thus inheritable, an experience-based product of evolution; an a posteriori of our breed, an a priori, a prerequisite for the cognitive mechanism of every individual. Transferred to the field of reflection, expectation and experience mean theories of prediction and proof, heuristics and logic, induction and deduction. Thus, there is, here at least, an overvaluation of the half circle of logic. For logic confines itself to the deductive inferences of proof theory. Induction, on the contrary, is a matter of heuristics, prediction theory (for this compare Fig. 29).

The dynamics of theories reaches a result in agreement with this. Erhard Oeser shows "that historically fixing the induction problem on Hume is not justified, systematic separation of proof and discovery is wrong and the understanding of classical induction problems in modern epistemology suffers from a fundamental confusion, namely of the inductive method with the knowledge-extending conclusions of propositional logic"(93). Investigation of the theory and history of our gaining of knowledge are confirmed in an astonishing manner.

Certainly, so Oeser(94) recapitulates, induction was understood by Aristotle as the opposite of deductive, binding inference. Whewell relies on the same insight, about a century after Hume's arguments. However, modern science has forgotten heuristics. It was not exact enough. As a result, science has fallen into the error that somewhere or other there must be an absolute certainty or truth, from which all other truths would strictly follow. Perhaps there is such a truth, but not for the cognitive apparatus of mortals.

The reasonableness of heuristics

How, then, did zoologists overcome the problem of their signally fatal error with the swans and what consequences did they draw from the establishment of an obviously misleading method of blind conjecture? The logician may well be disgusted. Zoologists never noticed this fact as a problem, and did what they have done successfully in a hundred thousand other similar cases: they extended the definition of swan by a colour and drew no consequences at all, except for satisfaction at having restored order amongst the relatives of the goose(95).

What entitles us to establish such generalisations? The answer, once more, is trivial: practical life. What colour feature should they have expected, that of the guinea fowl, the peacock butterfly or the violet? What does the logician expect to find inside when he sees his morning tram arriving? The characteristics of a holiday resort, of Vesuvius or of the Milky Way? He can expect only what he already knows of it, as he must, or else he would not be able to recognise his tram.

How can we convince someone who believes that the value of his knowledge depends on binding inferences? Wolfgang Stegmüller, in alluding to M. Black, has presented this situation so convincingly(96), though in reverse, that I shall repeat it transposed: How can one wean the inductive expert from his attitude? (that makes him use a rule about induction which we would all discard as absurd). Logically there is no way to undermine his conviction that what he follows are indeed valid induction rules. One may, perhaps, try to entice him away from this attitude by persuasive arguments. Still, we must suppose that reproaches of the kind that begin with, say, "you must be mad . . .!" will not dissuade him from his point of view! "Can one", Stegmüller further asks, "get through with an argument from success?" Here our answer, again transposed, is: yes, certainly; biologists will accommodate the next hundred thousand species to be discovered(97) without any problem, the two million known species will continue to solve their heuristic adaptation, and 2,000 million human beings will quite satisfactorily solve their heuristic life tasks; and we do not doubt that even logicians will recognise their means of transport despite their possible claim not to possess any binding foresight.

The biological reasonableness of this heuristic is guided ratiomorphically and certainly has its origin in its life-maintaining function; in the drive as well as in the foundation. Drive is the seeking mechanism of creative evolution. We know it from mutations, through association right up to the endogenous search mechanisms which themselves, according to Lorenz, Hassenstein and Eibl-Eibesfeldt, lead in turn from simple movement through restiveness and appetitive behaviour up to play, explorative behaviour and as far as research(98). Reflectively, it extends from fear to hope, neutrally it is called curiosity in the forms of expectation, foresight, idea and hypothesis. Initially, nothing more need be expected than "that something will turn up". For nothing would be more deadly to the learning process than for nothing to turn up. We shall come back to this when summing up (Fig. 58).

The argument for the specific expectation that something similar to previous experience will turn up is twofold. It is admirable that Stegmüller already had a presentiment of it when he says: "A way out from the dilemma" of inductive logic "would be found if we could rest on a non-logical principle, namely a synthetic statement about the world, perhaps on a uniformity principle which asserts that regularities observed in the past will hold in the future too"(99). We shall rest precisely on this; indeed all the possibilities of living organisms depend on this principle of constant expectation. It resides in the redundancy principle of this world and in the maintenance of the regularities that have arisen in it. One form is the actuality principle, whose application by Kant and Laplace, by Lamarck, Lyell and Darwin first made evolution conceivable(100). How absurd if we were to expect that tomorrow our world were to follow other laws than those it followed yesterday!

There is, however, another and more direct side to the argument. "The strategy of genesis" has the consequence that the "classification of living creatures" copies step by step the order in nature(101). A bare mention of our preliminary work on this must suffice here. This representing is itself the learning process, and it leads, as Lorenz, Von Holst and Tinbergen showed(102), to senses, data processors and apparatus for world pictures, which in turn are most finely tuned to what this world has to communicate in constant regularity. The story of all living creatures is selected for the further utility of experience undergone; every culture, every hypothesis, as Oeser shows, has such a history(103).

The aim of this method, the cause why it has survived, the purpose of all drive from simple movement up to research and the reason for all the imitation of nature up to the a priori of our ratiomorphic world insight is again survival. On reflection, this is the same urge to optimise orientation, foresight and understanding of this world, with the hope of peace, certainty, order and right.

It is amazing what deductive sciences have achieved; but dreadful that they make whole cultures think, even if unintentionally, that it is through them alone that one can gain any truth or certainty about the world.

The reasonableness of preconditions

Among ornithological riddles in logic that attract attention, Hempel's raven paradox is of some interest to us. Briefly it is this: "the statement that 'all ravens are black' is logically equivalent to saying 'all that is not black is not a raven' and this latter statement may be confirmed by something that is not black and not a raven, e.g. by a piece of white paper. However, since logically equivalent hypotheses are confirmed or shaken by the very same data, we should have to make the nonsensical statement that a piece of white paper confirms the claim that all ravens are black"(104). If we are speaking of our world, this shows that information derived from experience must not be considered in isolation. "A knee, else naught" we know from the poet Christian Morgenstern and at best from Hieronymus Bosch as well. The real world, on the contrary, contains constant interdependencies, without which we cannot imagine anything, let alone think(105). Things not only have characteristics but also their order and place in the world.

Only with this proviso can we understand the next venerable problem, namely that of concept formation. The evasive reaction of the paramecium, with which we are already acquainted, must be programmed with a "view" to obstacles, not to mating partners; the piercing instinct of the tick to bloodsucking, not to flight. Abstraction fulfils its life-maintaining aims only within given areas of validity, but then it fulfils them excellently. Within their particular programmes the definitions abstracted, "obstacle" or "mammal", acquire meaning, and fulfil their task. Then, the most reliable and most constant feature of the vital circumstances in question can be abstracted. The hierarchy of instinct programmes (Fig. 37) will show us how high in the stratified structure of presuppositions the genetic working instructions can reach.

The processes of generalisation which, in the form of simplifications, associations, or abstractions, extend to mental pictures and concepts, definitions and statements will be discussed in Chapter 3. They belong to the "hypothesis of the comparable". Here it is only the system of their pre-suppositions that is important, and we recall that individual learning continues to build on the same pre-suppositions. The conditioned eyelid reflex does not cause any other bodily part to twitch, but is conducted by the unconditioned route to the eyelid. In the case of conditioned food appetite in the dog, no other gland is stimulated apart from the salivary gland. Of course, associations extending into consciousness are formed precisely with the same content of pre-suppositions. What use would be the correlation, developed between the starting and control lights, if it applied not to the particular experimental arrangement but to the lights in a lift or in traffic? What would be the point of the raven's external features or indeed its black colour, if they did not apply to the raven, and continued as such within the genus Corvus and then in birds, vertebrates, Metazoa and organisms; with all the relevant background knowledge and all the pertinent hierarchical preconditions?

The more differentiated the objects are to which organisms react purposefully, or as we put it, which they understand, the more comprehensive do the preconditions of that background knowledge become. The spheres of validity, within which the heuristic sector of knowledge-acquiring reason can match its life-maintaining functions, must become correspondingly sharper and more differentiated.

The necessity of the cycle

To sum up, the proof-theory of logic can achieve no more in solving the problems of life than the foundation for expectation, that heuristics can achieve alongside it. However, dazzled by the imagined certainty that we find possible in the artificially isolated sector of deductive reflection, we have forgotten that the latter receives content and life only through the inductive sector.

"Thus Cicero already points out", says Oeser, "that the Stoics did indeed further develop Aristotle's syllogistics as an "ars iudicandi" whilst the Topic, as "ars inveniendi", the teaching of premisses and rules of inquiry, was left unconsidered. However, according to Cicero, Topic is not only useful but also "ordine naturae certe prior" (106).

Ramond Lull, the founder of search rules, even regards them as more important; Bolzano takes Topic and heuristics as equals and, in his epistemology, continues with it as the art of discovery. Logicians paid attention to inductive concept formation only, as we remember, as far as Frege, and more recently only Polya has come closer to heuristics(107). In modern introductions to philosophy and psychology, in logic or epistemology, it no longer occurs, not even as a catchword(108). In the natural sciences, it scarcely occurs at all.

How very reasonably the ratiomorphic apparatus continues to guide us, like that inborn teacher, sound common sense, the heuristics of our cognitive process, as we rush from discovery to discovery. Although heuristics may be undiscovered, forgotten or even denied, there is no doubt that it represents the indispensable half of our knowledge-gaining reason in the cycle of expectation and experience.

SENSE AND NONSENSE IN PROBABILITY EXPECTATION

One does not readily profess oneself a friend of prejudice. Too often, judgment in advance has proved simply to be wrong. What we value is a cautious, considered judgment by reason, with all its qualities bordering on wisdom. Still, we should be quite inviable if we did not continually guide every tiny section of our actions, our constant and scarcely perceptible decisions, by judgments which, in spite of the limits of the known, reach beyond into the as yet unknown. We have already met the biological roots of this driving force and, in man, these guiding brain centres have been disclosed by clinical psychiatry(109).

When we approach some puzzling and suspicious object very carefully, even with scepticism and mistrust, our opinion turns out to be permeated by completely unimagined expectations, by completely unconsidered prejudices which may change but can never be absent, because otherwise our actions would lose everything, drive, motive and hence regulation and guidance at once.

What is indispensable in prejudgment

Actually, prejudgment at all levels of our structure is as indispensable as it is long established, to relieve the organism of decisions that could never be accomplished by trial and error or, at least, not correctly and in time. Judgment as to what material our body should produce and at what place, which muscle (perhaps on slipping off a step on the stairs) to contract and how strongly, what changes in retinal image to make when something rushes at us(110), and after how many disappointments of an expectation we must reckon with chance, or after how many confirmations with the same reaction of things, all this must be taken from us rational beings, if we are to survive. Even in the mental field, prejudgment of "attitudes" is necessary. As Hubert Rohracher says, "man would otherwise be constantly exposed to perplexity and uncertainty, he would have constantly to engage in long, laborious and difficult deliberations, and he would not know where he stood in spiritual reality"(111). Indeed, even society and civilisation, which organise our world, are, as Berger and Luckman showed, themselves made up of a world of prejudgments that the individual could never check(112).

Prejudgments, therefore, are doubtless a prerequisite of our existence. Wherever their advance decisions hit home with some probability, more readily than a random search could, where it is successful, namely where it protects conditions of life and survival, there it is acting sensibly. This is the aim of prejudgment. Where it hits home with certainty, it acts as if with foresight, knowledge and wisdom. Where it cannot succeed, it is like stupidity, or sheer nonsense.

The success of prejudgment

Prejudgment is thus indispensable as a motive for finding decisions. The success of prejudgment, however, depends on learning and experience and hence on the acquisition and possession of knowledge. Knowledge gaining or adaptability originate first in the uncertain trials of mutation in individuals, in systems of society, and secondly in selection, the selection of success by the superior conditions prevailing. Possession of knowledge depends on the preservation of success by the molecular and individual memories and by tradition(113) in the memory of civilisations.

What may seem strange in this process is the necessity for chance. However, we must bear in mind that where we already know something, decision finding by experience must be superior to that by chance; conversely, where we cannot know anything, the enquiring chance judgment will have a reliable chance of success. For judgment from experience can always, fortuitously, exclude the target field from the field of search and thereby any chance of success. There are numerous examples that show how often it has been precisely outsiders who have initiated decisive discoveries. As Thomas Kuhn shows, the scientific revolution must always be seen as a turning away from traditional knowledge(114). Here we must take it as certain that solution finding by the individual and his society will alike succeed more reliably by chance exploration wherever nothing can be known. We recall that evolution has retained this necessary chance in the form of molecular chance for learning by the gene, but for the learning by the brain, as a consequence of long causal chains through thousands of neurone switching points(115), as we shall see.

The lack of success of so-called magic thinking or prejudiced attempts at solution is well known, not only from everyday life, but also from many experiments in psychology and in behaviour studies(116). Among the simplest examples, perhaps, is the problem of joining the nine points in Fig. 15, using only four straight lines, which can be solved only if one sheds the prejudice that the figure suggests(117). Likewise, all creative intuition appears ultimately to be a chance achievement of the unconscious. Here unprejudiced consciousness must leave the field to chance, as both Manfred Eigen and Ruthild Winkler show(118). It is one of the most fundamental prerequisites of creative innovative (fulgurant)(119) evolution that it cannot do without chance, as the generator of variability. This basic principle has been made particularly clear by these two authors(120).

The success of chance

That the success of chance extends so uniformly from solution searching by molecules up to that by creative thought depends on the constancy of both external and internal causes. The external cause rests on the same probability principle governing the chance of success that functions at all levels of biological learning; this is a functional analogy(121). The internal cause, on the other hand, rests on the genetic handing down of the same processing mechanisms; for, in the learning algorithm of molecules, we have recognised the switching instructor, in both cases the prerequisite for imprinting conditioned reflexes and so on. Each newly created learning layer, as long as it has to function, makes the maintenance of all deeper ones irreplaceable, just as the upper stories of a building require maintenance of the lower ones until, finally, all the lower layers become the prerequisite of ratiomorphic reason and this in turn the teacher of pure reason. The biologists calls such stable genetic building instructions homologies, in contradistinction to analogies, which follow the selection instructions of external conditions, the result of obeying selection instructions, that arise in the organisms themselves and are co-inherited(122). We shall come back to these concepts in Chapter 3. The biologist knows many such functional analogies on a homologous basis as so-called homoiologies. These are functionally equivalent adaptations of related structures: for example, the functional analogy of the fin structures of quadrupeds that have returned to the sea(123).

The success of selection

If we recognise the success of a uniform chance generator by homoiology, we see at once that the success of selection depends on the available range of selection. In the mechanism of evolution, the maintaining of chance is essential, but encountering the necessities of selection is, time and again, fortuitous. The selection conditions, which an organism comes across unforeseen, in no way require it accurately to reflect the laws of the world, but only that tiny segment of them, which by chance acquires vital significance for it. All conditions lying outside this selection range, however important the insights extractable from them might be, remain unconsidered and cannot even be ratiomorphically conjectured. That is why world views extractable from the selection range will be right within that range, but extrapolated outside it, most probably wrong(124). This prepares the solution of the reality problem.

The success of the probability calculus

The success of the probability algorithm has now become firmly established within the selection range, which the nearly endless chain of our ancestors has traversed. It asserts, according to our rational copy of it, that we approach closer to truth if we start from hypotheses of whatever probability, provided we test any expectation from it by experience and insert it for the sake of gradual improvement in the next hypothesis at each stage.

We must, therefore, always start from a position that indeed contains some expectation but very little experience.

In ordinary usage, we call this an unestablished, purely subjective prejudice. We can also observe such an attitude in ourselves, wherever the wish is great but the knowledge vanishingly small. It is quite moving to see how subjective emotional expectation judges probabilities with a feeling bordering on certainty, when, objectively, one cannot have the slightest idea of them. For example, one builds on the notion of finding a misplaced key perhaps in the house, or even in a particular drawer with a probability of near certainty, although one is not even certain of having mislaid it there. Or, conversely, one will exclude the possible occurrence of an event, perhaps an accident, with a probability near to subjective certainty, although one can have no experience at all of the intended event. "Nothing will happen to me", so runs this kind of judgment about a completely unknown domain of possibilities. On the other hand, one who may be considered an expert on what is in such a domain, will feel justified in asking the question which we formulate after Stegmüller(125): "Are you mad?" What is in control here, we know as life function, appetite, wish or hope. However saddening the fact that Stegmüller's question may often be justified, this control which continually motivates the whole world of organisms to probability hypotheses about the unknown, is a principle of all creative evolution. It is the endogenous mechanism of heuristics. Symmetrically with subjective probability of logic, as with Finetti, Ramsey or von Kutschera, we may call it the subjective and irrational probability of heuristics(126). Although it is unprotected against humbug, be it that of gods, demagogues, society or education, for our peculiar rationality it always stands at the beginning of every path to wisdom.

The necessity of probability calculus

Now, the path that we can follow from subjective to objective and rational probability(127), corresponds to the gain in knowledge, to optimising the hypothesis from the constant interplay of expectation and experience. In this, the degree of objectivity and reason is represented biologically as an endless continuum (Fig. 16). Therefore, instead of asking how objective a probability must be to be objective, or just how reasonable it can become(128), heuristic reason seems to operate with degrees of certainty, which we have already met. These may border on absolute certainty but can never quite attain it. Not even the most objective and most rational probability ever suffices for binding, rational inference but it usually does for probable inference by living creatures seeking direction. With this we are close to a first solution of the problem of induction.

Now the degree of certainty which we think we possess regarding phenomena of this world, can through conscious reflection react on our probability expectation. For example, deductively, from the geometry of dice, we expect that each of its sides, after very many throws, will make up exactly one sixth of all results. If this does not happen, then we would rather doubt the precision of the shape of the die or the position of its centre of gravity, than our expectation.

Conversely, we would sooner trust the movement of the sun rather than that of our watch, although we can rationalise the laws of astrophysics and we know that in the course of 6 or 8 thousand million years, the sun will one day no longer rise (because it will have become a giant star and have burned up the earth)(129). The reason of the living, however, reckons with certainties within the measures of life spans; although we can perhaps calculate from the random thermal motion of molecules that once in a while the dice will not fall at all but could fly away (namely, as we find with Bernhard Bavink, if the random motions of the molecules by chance all lead in the same direction)(130); although we may calculate, according to Roman Sexl, that even in physically ideal billiards, the seventh ball need no longer strike the eighth (because the surface molecules' positional indeterminacy raised to the eighth power is as big as the billiard ball itself)(131). However, in spite of the subjective feeling of absolute certainty, the mechanism of the ratiomorphic hypothesis dictates action according to what is most probable at the time and treatment of this action as justified, only until sufficient failures of an expectation force us to give up the hypothesis and formulate a new one, which in turn can direct action based on presumed certainty. It is as though chance, rooted in microphysics, had its most complicated consequences in a scepticism about its establishment in mutations and in the freedom of creativity due to the long chains of switches in the brain, in a constant residue of uncertainty regarding the ratiomorphic processing of our expectations about this world.

The relation to rational theory

In distinct contrast to the ratiomorphic algorithm of the apparently true, stands that of the rational, consciously reflected theory of knowledge gaining. It seems to us as if it were a rationally ungrounded belief in an ultimate reason or purpose, in some place of absolute certainty, that has made philosophers adopt a viewpoint foreign to the nature of our world. In Chapters 4 and 5 we shall indicate the causes of this.

Here it is already obvious that reflective reason was not interested in statistical but in binding certainty. Naturally, this was to be found only in deductive, mathematical and logical formulations. Consequently, the "science of right thinking" retreated to a science of correct deduction, the theory of science to a theory of correct proof, and heuristics to negative heuristics(132). Even Carnap's whole continuum of "inductive" methods is a test in a deductive direction(133); perhaps of the probability of a hypothesis for the data supporting it. Popper's quasi-induction, too, in its concrete steps runs in this direction. Science has excluded the inductive sector of the cognitive process, failing to notice its preconscious activity; so far, science could do this without penalty because that process functions outstandingly well. When Popper asserts "there is no induction"(134), he could only have meant induction as a method of logic; for precisely this has withdrawn from heuristics.

Thus, the theory of science, by its retreat to formal truth and the claim to absolute certainty, has denied the building, creative half of the knowledge gaining process, namely positive heuristics. However, it is this that we are now expounding as a principle, as a mechanism of evolution; and we have derived from it the same positive expectation theory that Ernst Mach and Whewell had in mind and as it has been reconstructed, especially by Erhard Oeser, from the "dynamics of systems of empirical science"(135).

Solving some riddles of reason

The central question that may be answered from our evolutionary standpoint is that of the a priori, first with reference to probability. For, as Kant formulated most fully, if our reason is so organised that it must possess certain judgments in advance in order to comprehend this world, then two things follow. First, it may be shown that the a priori cannot be based on ratio alone, because they are its prerequisite. Secondly, the question remains as to how the a priori came into reason. Vollmer(136), along with Stegmüller, called this "a fateful question of philosophy". Our answer is: the a priori of pure reason have entered into it through the evolution of the ratiomorphic apparatus. They are a posteriori, hence experiential products of the knowledge-acquiring mechanism of the living.

The "hypothesis of the apparently true" contains the biological background of Kant's a priori of modality, the expectation that we can presuppose this world to contain "possibility-impossibility, existence-non-existence, necessity-fortuitousness" as Kant termed them(137). This hypothesis, too, could be successful only because it has reproduced the principles of redundancy and interdependence, contained in this world.

This provides the first solution of the homology problem, a first reason why we can grasp related similarities of organisms. More comprehensively, the problem of "categories", already known to Aristotle and a problem that is permeating the whole of western philosophy, is beginning to solve itself. We can state this because we foresee that in the next three chapters we shall solve the other three categories of Kantian a priori. We can be brief here, for we shall summarise all this in Chapter 6.

Let us recall that this possible solution was first discovered by Konrad Lorenz. As Donald Campbell recognised, it became the basis of evolutionary epistemology and consequently the start of all thinking, as philosophically grounded by Gerhard Vollmer, a "third Copernican turning point" of scientific thought about the origin of thinking(138). It became the basis of all that we have here established biologically.

Finally, it should be noted that, for Kant, the categories of modality came at the end of his analytical procedure; but for us at the beginning of our synthetic one(139). For here, the "hypothesis of the apparently true" is shown to be the precondition for the formation of the remaining hypotheses. This agreement of their discovery by the two opposite processes of investigation strikingly confirms their concordant position, and the analytical-synthetic, or deductive-inductive research procedures. We owe this insight to our seminar with Konrad Lorenz and

Erhard Oeser. Once more we can confirm Kant's statement that "the categories of modality have this special feature: they do not in the least enlarge the concept to which they are joined as the predicate as determination of the object, but only express the relation with the cognitive capacity"(140). The hypothesis of the apparently true is simply the prerequisite for the cognitive process as such.

Everything that the inherited hypothesis of the apparently true contains, therefore, counts amongst what has sense in prejudgment.

The nonsense in prejudgment

As we can already foresee, nonsense in prejudgement must be expected wherever a judgment seeks to go outside the domain of the experience that has developed it, by extrapolation. Here begins the nonsense of "learned molecules", of "learned switching paths" and even of "learned academies".

We should remind ourselves that the ratiomorphic algorithm leads to regularities, as we say, being regarded much more quickly as certain than would appear justified on reflection. By contrast, no organ seems to have been developed that is able to detect chance directly. This structure of the innate teacher must have been fully adapted within the framework of life problems in animals. It was enough, indeed wiser, to accept possible necessities as true as quickly as possible. There was no need to store everything that was fortuitous. As far as practicable, it was to be left outside the scope of attention and registration and was in no case incorporated in the molecular memory. However, within the framework of the life problems in man, which are greatly extended by his reflective reason, it looks strange that ordinary common sense has the greatest difficulty in recognising chance as such in sequences of events.

Conversely, if our reflective reason attempts to switch off the ratiomorphic instructor, to restrict solution finding to the acquired life range, as it were, then two fundamental errors occur. Chance can, quite rationally, be taken for necessity and necessity for chance. Both can be justified in the imagined sphere of rationalisation. Both would be catastrophic to the immediate range of everyday life-maintaining decisions.

It is quite obvious that for an enlarged environment for which it was not created, the ratiomorphic teacher will no longer instruct appropriately; and that rational reflection which is devised for the additional domain can become wrong for the starting area.

The limits of the domain of selection

In inherited programmes, the instructor's range of experience still coincides with the domain of selection. In that domain, the programmes must contain the correct advance judgments because they are a product of selection. Outside the domain, the prejudgments will be completely wrong; and this will be the more probable the further one extrapolates and the further one departs from the test area.

It almost seems wise that the prejudgment of molecules should guide the paramecium in the direction of acids in water because these indicate the presence of bacteria which are their food. The same prejudgment, however, may lead all paramecia to death in the presence of strong acids in the experiment(141). No less reasonable is the prejudgment for the switching of many soil animals which, under dry conditions, directs them downwards into moister soil. However, if the soil animal collector puts a soil sample in a funnel (as shown in Fig. 17) and proceeds to dry it from above, then all the migrating animals fall through the base of the funnel to no less certain death in the collecting vessel(142). Conditions of that sort, and we know of similar ones with humans(143), never figured in the selection range of those advance prejudgments, nor indeed could they have been expected.

Nevertheless, the limits of the neighbour's capacity for judgment are already exploited by nature for her own advantage. We recall the imprinting of emerging ants, which establishes for them the image of their species in the ant which helps in this process. The slave-owning ants have promptly utilised that weak point in the genetic prejudgment so as to obtain the life-long service of slaves through the modest effort of obstetrics(144). After all, a good part of that cycle of consumption that we call the economy of nature, lives off the deficiencies in prejudgment of the neighbour. Judgments in advance, however, are necessary, and along with the wrong judgments that they entail, they are tolerated so long as their success exceeds that of other prejudgments as well as that of indiscriminate searching.

At first the associative, individual learning of prejudgments from conditioned reflexes seems somewhat less risky. However, we are all familiar with the difficulty of shedding attitudes that have been acquired associatively, as Hubert Rohracher shows(145), even if life makes it clear to us that they are wrong. The evolution of mankind, however, goes further still through that fundamental change which means that the whole of individual learning need no longer be lost for ever in the grave. This second evolution has developed the inheritance of individually acquired learning; through imitation, speech and writing. This again causes prejudgments to be passed on — from man to man, and from generation to generation, all of which, in turn, possess the indispensable advantage of advance judgment within the sphere of selection; while outside it they may again become sheer inherited nonsense.

The shrinking of the selection domain

With the gains by civilisations, as they have now developed in consequence of the second evolution, with masses, empires and power blocks, new systems of conditions, and of selection also, have arisen. Since direct cannibalism among groups of Homo sapiens(146) has become very rare and the tribe has extended the protection of separate individuals as far as territorial borders, a shrinkage of the selection area begins within the system. Fostered by the unbridled associative delight of consciousness, attitudes that are quite divorced from reality can accumulate indiscriminately. If it be a privilege of life to have developed extrapolated nonsense, then it is the privilege of man to believe in pure nonsense(147).

Of all the nonsense that human reflection has brought about, the confusion of probability is a part. Here, through the power anchored in tradition and in defiance of all ratiomorphic warning, chance can be mistaken for necessity and conversely; which would have at once destroyed any other species.

Since the earliest cultures, the fortuitous has been taken for legality. From the divining of entrails and the phantasies of astrology(148), the nonsense drags on to dream books, palmistry(149), then on the one hand to harmless superstition about lead pouring or being afraid of a black cat crossing one's path, but on the other to the pernicious demand on truth of the kind of humbug, according to which not only fates are sealed and battles plotted, but whole peoples have been and are being exterminated.

The necessity for natural laws has no less been taken as fortuitous; since classical philosophy, both the reality of matter and the reality of the spirit have been doubted. Cultures have been eliminated in the name of purpose, because they appeared to be fortuitous and therefore pointless. Over the two thousand years of our intellectual history there have been not only implacable disputes between idealists and materialists, but these half-truths have hardened through their demands on truth and power from supposed truth demands into ideologies whose contradictions today endanger the entire world. The selection available to us has grown into collective selection and into collective responsibility for collective nonsense, which against any humanity must carry off everybody, stupid as well as wise. All this is part of the nonsense of prejudgment, wherever the hypothesis of the apparently true exceeds its tested limits.

Truth and perception, understanding and sense, certainty and deceit accompany the whole history of human reason; we have found them, as antagonists, along the entire pathway to knowledge. In the first as well as the second evolution, they are the antipodes in the most fundamental processing part of our total world view, namely in the hypothesis of the apparently true.

CHAPTER 3
THE HYPOTHESIS OF THE COMPARABLE

"Thus we ascribe to nature . . . a consequentiality, a rule by which we presuppose it will proceed . . . and a transformation which constantly changes . . . the parts named under the type."
Johann Wolfgang von Goethe

"Man is 'provided with certain natural beliefs that are true' because 'certain uniformities ... prevail throughout the universe, and the reasoning mind is itself a product of this universe'."
C. S. Peirce cited by Noam Chomsky(1)

Who is to decide when the unlike is like or the same different; a neighbour, inspiration, experience? How could one trust any one of these, seeing that they contradict each other all along as to what is the same from birth onwards, before God or before a judge? Was it not some higher authority that always decided, war in the sense of inequality but revolution in the sense of the equality of man; reformation versus counter-reformation, nobility against the proletariat, employer versus trade union? To which authority should we appeal, since all higher ones too have been contradicting each other from time immemorial, from the contradictions of the world creator and his demiurges(2), to the contradictions of metaphysical systems and their ideologies, the demagogues and idelogues of our day. Is it not simply the intervention of the more powerful, at any given time, that decides on certainty?

Like and unlike

amongst things, people, ideas or visions, are the next chief characters of the scenario in which act by act life and consciousness, thinking and a world view could arise. The dialogue between them likewise extends throughout the whole confused disorder of our ever unfinished history.

Where, then, lies the foundation that enables us to grasp this confusion of things, and their states and events? Do we assume that we can grasp it because it is real, or do we assume that it is real because we can grasp it? As before, a world of Platonic ideas(3) contrasts with a world of names, reason against experience, mind against matter(4); and it can be reckoned as only probable that without an ordered world of objects, neither would a certainty of probable truth make sense, nor the question of its causes and purposes have any content. Therefore, the hypothesis of the comparable has to be joined to the hypothesis of probability. That is why the cognitive process of the living has long since developed it; again this has become entangled with consciousness and we have to carry on with it.

WHEN UNLIKE WOULD BE LIKE

The very word "compare" embodies the problem; namely putting on a par with or making equal what is unequal(5). For the question arises at once what entitles us to make equal, and what is gained or lost by doing so? For example, in making men equal, the most essential feature can be lost, namely what is most peculiarly human. The uniqueness of each could vanish, the irreplaceable in his unitary individuality.

When like is never the same

Yet we can discourse on uniqueness only in a language consisting exclusively of things made like. We except only proper names, and the proper names of some men's works. However, even if we describe Michelangelo Buonarroti or the Sistine Chapel, we have to mention names or properties which, though isolated, occur in many renaissance men or in roof paintings of their time. Even patent applications for some unique discovery must be comprehensible in terms of tubes, valves and regulators, which have long been common and widespread in our world. How else could we make ourselves understood? All language must depend on things made like. That the complete identity of several actual objects in every respect is impossible, however, is a basic thought of many philosophical systems(6). In fact, even what is like can never be the same. Indeed, in the same motor car of the same series, or in the same egg of the same hen, two like atoms are never the same. If I have the same sentence set a second time, although the symbols are the same, they consist of other molecules of the same printing ink in other places of the same page; like things can never be the same. If the reader thinks he will be still the same in a few hours' time, nevertheless, thousands of his cells will have died off and will have been replaced by others(7). Greek philosophy already knew that you cannot step into the same river twice(8). Just as in a wave, all parts are changing as long as it progresses. What, then, entitles us to set equal what is never like?

When there are no limits to what is similar

The situation becomes still more difficult if we recognise that the similar has no boundaries. For whence would boundaries arise, say, between dust, sand, gravel, stones, rocks; or those between huts, houses, castles, palaces? How many grains make a heap? asks Hassenstein(9). Is it not simply artificial boundaries that we set, following a prescription of our thinking? Indeed, science must not hesitate to define the limits of its concepts(10). Generally, however, it is not at all clear which of the many measures containing a similarity will be the one preferred by the definition. Already five figures with only two variable quantities, as shown in Fig. 18, permit ten different, well-defined units without it being possible to distinguish between the right and wrong boundaries(11). However, natural objects of which we are aware always contain so many recognisable quantities that the number of mutually conflicting boundaries is enormous. It would be absurd if we were to draw sharp boundaries when defining hills and mountains, shrubs and trees, or ships and boats. This has a further consequence:

When there is far too much unlike

Where we intended to solve the induction problem as a probability phenomenon, are we not again led into the insoluble, since we cannot know from which of the many special quantities we can infer to the general? What if the special allowed arbitrarily many forms of the general to be set up? Is not then the observation of the special the final solution?

This hope cannot be fulfilled, either because there is nearly always far too much of the unlike. Just walking through a busy large town brings ten people per second before our eyes, which makes more than 100,000 in three hours. How many can we describe, in retrospect? We have in fact glanced at many details of every individual, but have become really aware of few and have got to know even fewer(12). Only the crowd and a few of its notable features stay in the memory. What sticks in the memory is merely what is common to them, the general feature of many rain-soaked, dust-enveloped or lightly dressed people. The general, the like simply had to be formed. The excess of individual detail we cannot even grasp.

When the actually like has no content

However, the worst feature of the unlike is that the genuinely like would no longer have any content. Every difference would have to be removed. Indeed, we do not know how to compare qualities, but measurements and mathematics show us that quantities may be unobjectionably compared. However, Konrad Lorenz(13) says, "the counting machines used in our extensive quantifications, work somewhat like a bucket dredger, adding a small bucketful of something or other to what has gone before. Its work is correct and consistent only when it idles, counting only the returns of its single bucket, the unit. The way in which we allow this machine to intervene in the heterogeneous material of extra-subjective reality, the absolute truth of its statements is lost at once." Thus, the like is certain only where it no longer has any content.

When, therefore, could the eternally unlike be made equal, or be compared? How could the general consist of the special, if the multiplicity of qualities remains undetermined, but its quantity loses content as precision grows?

In spite of all this indeterminacy, we find our way quite adequately in this world, having made the countless objects and states that we perceive equivalent in all languages by the hundreds of thousands of our concepts; and without being aware of how we proceed. In the same way, the classifiers have coordinated the countless individuals of living organisms with the concepts of two million species in a hierarchical concept system of over five hundred thousand affinity groups; they correspond to actual similarities in such a way that this "natural system" establishes one of man's most profound insights: namely the knowledge of his own descent.

However uncertain our conscious reason may be regarding the process of comparison, unconscious reason must possess a method that is entirely successful at comparing; this we must now seek out.

EXPECTATIONS IN THE PROCESSING OF DATA BY LIVING CREATURES

None of the lower organisms has wanted to learn anything; neither the coli bacterium, nor the paramecium nor soil animals nor ticks. We are sure that they had to be compelled to learn(14). Therefore, whatever the genetic memory has learned as to structure and instructions for its organism must have developed under the conditions prevailing when its ancestors were alive. The chance generator of mutations has created the variability and selection chose the most useful at any given time.

Such a mechanism in turn presupposes that in the habitat of organisms there is something to be learned at all. As to the hypothesis of comparability, we must start from this question:

What is there to be learned in the way of similarities?

What one can learn from the world is its order. Disorder can certainly be produced, indeed, it must be, as we saw; but there is nothing to be learned from chaos. The most basic thing about order is the coincidence of states or events. That is to say, most things occur with great regularity only together, in succession, or in connection with certain other things. In the case of man, this is so self-evident that often he no longer thinks about the fact that thunder and lightning coincide or that a fall of rock is followed by a rumbling noise, or that fruits are associated only with plants. Thus, inheritance programmes of organisms depend on an

Abstraction of coincidences in nature

The turn-round reaction of the paramecium (Fig. 4) extracts from the welter of unknown properties of the obstacles, the coincidence of solid surface, resting position and bounded extent. It somehow disregards all the other properties. The same inherited instinct of the tick extracts the coincidence of butyric acid and the temperature of 37°C from the many unknown properties of the mammal.

If such an extractive performance by selection is translated into our rational mode of expression, then we speak of the abstraction of the essential; or one might say that in observing a characteristic, the mechanism includes the expectation of meeting with other but now quite specific features. Accordingly, we shall find this expectation again in our consciousness.

How to separate like from unlike

What is decisive for the biological mechanism that has to extract the order of similarities from this world, is the question of how the like can be separated from the unlike. Here, too, the reason of the living, the algorithm of data processing is convincingly simple and amazingly accurate in aim. The method consists in removing from the abundance of properties those that most regularly coincide. This amounts to the task of separating the constant from the variable, the presumably essential from the fortuitous; and this task must be fulfilled because determining the constant and the essential is of vital significance as it substantially increases the probability of accurate foresight, and hence precision of the judgment in advance.

Thus, the turn-round reaction does not extract, say, any material features of obstacles, such as the silicate or cellulose contents, because they can change; neither is the surface structure incorporated in the programme because this also varies widely from sand grains to algal threads. It is rather the firmness, size and resting position that are taken into the programme, as in most actual obstacles these properties coincide. This mode of operation is still more distinct in the inheritance programme of the tick. Every student knows how many characteristics there are from which to diagnose a mammal; hairs, milk glands and the most complicated structural characteristics of the internal parts. He knows that the most distinctive characteristics are those of the internal anatomy; the organisation of the heart chambers, vessels, kidneys, and so on. Even more clearly he has in mind the many other characteristics that do not at all coincide in all mammals; such as claws or hooves, tusks or antlers(15). However, what does coincide in all mammals is the body temperature and the smell of butyric acid which is a fermentation product of the tallow secretion from the skin glands. In fact, there is no simpler or more reliable definition of land mammals than by temperature and butyric acid. And it would be almost impossible for it to be an error of the advance judgment that this coincidence should further coincide with the possession of a skin that the mouthparts are adequate to pierce, with hairs to which the climbing legs are adapted, and with blood which is indispensable to the metabolism of the tick(16).

Abstraction according to the level of coincidence

This is carried out by the IRM, the inborn release mechanisms, in the next higher levels of the inheritance programme(17). As soon as the sense organs supply as much data as our ear or even more our eye, stimulus filters must be built in, which are suitable for conducting only such information as is pertinent to the reaction in the pre-set programme. Thus, the female cicada, in order to exclude errors, recognises only the mating call of the male, out of all the many chirpings in the surroundings(18).

Wherever the sense data are complex and multiple, the filter of the IRM accepts the information with the highest degree of coincidence out of the abundance of possible variants. This is made especially clear in dummy experiments(19). For example, the robin is found to recognise its own species by reacting only to the red patch on the throat, of all the other variable details. For even the best dummy robins will not be recognised if the patch is missing, whereas a bundle of red feathers will be taken for a female (Fig. 19). Likewise with the chick of the herring gull, which prefers a red stick with a white ring which resembles the beak of the mother, to the mother dummy that is most similar in all characteristics(20).

This selection of characters according to their degree of coincidence becomes so significant that selection finally applies striking species-specific signals to the information carrier itself. Thus, the throats of many young birds are provided with highly specific, dazzling colour patterns, so as to ensure that their parents can feed the open mouths of the young with the maximum accuraccy (Fig. 20). The males are equally provided with mating signals and a whole series of warning, fears and flight signals(21) so as to be able to extract with certainty what is constant from what is variable. In this, selection proceeds so consistently that we can grasp how unambiguous and sure of aim prejudgments must be to preserve the species.

The perception of form

At still more complex levels of genetic programmes, such selection-tested advance judgments become pilots in the domains where we judge. Such, for example, are prejudgments that govern our perception of form. It is a peculiarity of our consciousness that they remain hidden to us wherever they correctly judge in advance; but where, beyond the range of their validity, they become contradictory, they become manifest as optical illusions.

Thus, in most cases with mutually moving parts of an image, it is correct to consider the smaller section as moving and the larger as at rest. For example, it is very reasonable that even a suckling shows defence reactions as soon as an advancing object, if only in a film projection, seems set to collide with it. This reaction can be of life-preserving importance; but again, only if it judges in advance. In retrospect, it is too late. Likewise, it is appropriate that whatever moves together is taken as being connected, what moves in space is taken as spatial and what is spatially distant must be expected to be reduced in size and allowance has to be made for the illusion that the apparently three-dimensional occurs in two dimensions on paper (Figs 21 and 22). Even the urge to complete in advance an incomplete perception is usually of vital importance. For a gazelle, it is equally important to supplement the sight of the tail of a lion with the whole form of a lion, since the occasional deception which this prejudgment brings with it is generally less dangerous than neglecting the advance judgment(22).

Indeed, the prejudgment of the advance appraisal of form only becomes absurd in areas of civilisation for which it was never selected, as will be shown in the third section of this chapter.

On effort and result

We shall presently examine more closely the fact that this extraction principle contains an economy principle as well. It is already clear at this stage that the accuracy of aim of an unavoidable judgment in advance, and hence the degree of probability of a correct foresight, must occupy a prime position in life's economy. For the right judgment in advance not only saves energy and the time of uncertain trials but it also often avoids the kind of dangerous risks inevitably linked with aimless experiments. Efforts and result are closely connected in all processes of living organisms.

Further, this explains why the construction of genetic programmes and the storing of data in them proceeds with extreme economy. Konrad Lorenz(23) said, "When one has seen how paramecia cleverly remain near nourishing bacterial colonies and how promptly a newly hatched turkey chick presses into the nearest cover at the sight of a flying predator bird, or how a young kestrel, coming into contact with water for the first time, bathes in it and then polishes its plumage as if it had been doing so already a thousand times, one is almost disappointed to find that", as we already know, "the paramecia merely orientate themselves according to acid concentration, the little turkey likewise seeks cover on seeing a large creeping fly on the white ceiling of the room, and a smooth slab of marble triggers off the same movements in young kestrels as does water."

Here, effort and gain are reflected once again. It is certainly difficult to imprint environmental images in the molecular memory. It may require a hundred thousand generations and a million years for an experience to become a genetic feature(24). It must be dangerous to programme details that may later change in the environment. For it is clear that for the molecular memory it is at least as arduous, time-consuming and risky to unlearn what is false as to learn what is true(25). The calculation of effort and result is a basic principle for the living. Even molecular learning is subject to it. Is individual learning free from it?

Abstraction in individual learning

We already know about the probability hypothesis of the living, that it is the slowness of molecular learning which must have promoted the development of individual learning. Doubtless, this learning process is speeded up by several orders of magnitude(26). Still, the old principle remains. Abstraction in individual learning is distinguished first by the mode of information storage; it now takes place in the brain instead of in the genome. For no one wanted to learn, and the principle was again based on the conditioned reflex, on coupling or association of a firmly established (unconditioned) programme and on the extraction of the most constant coincidences.

The classical example of Pavlov's dogs (Fig. 22, also cf. Figs 8 and 23) may be recalled. If the feeding bell for the animals was sounded repeatedly and constantly enough at the sight of food, in the end saliva flowed at the sound alone. Things are associated if they seem firmly connected; and they seem firmly connected if they constantly coincide. For only what is necessarily connected by the condition of nature will usually coincide. The extensive literature on learning processes(27) has long since established this. As we shall see later, only rational learning can succeed in taking the fortuitous as necessarily connected. This is another paradox of conscious reason.

That coincidences can be learned is shown, in principle, by the simple experimental conditions of training. Because, as in our conditioned eyelid reflex, dogs actually associate only what the experimenter quite deliberately combines; if he takes care to repeat the association again and again, without gap or inconsistency, without mistakes and without omissions. However, after a few coincidences of bell and food, the dog still does not associate anything. We know from all training, that the combination has to occur often, indeed very often(28), before an animal shows by its behaviour that the connection has been formed, the association has been fixed; until the animal has grasped it (Fig. 22), as we quite rightly say.

It is self-evident that the connection can be lost again. Stop ringing the bell for long enough and the association will be forgotten(29). The unfulfilled expectation will even more effectively dissolve the hypothetical connection; if, for example, the food bell sounds, the saliva flows and the dog, excited at expecting food, finds that no food appears during the sounding of the bell. Repeated disappointment has the strongest effect(30). It may result not merely in embarrassment and irritation but can also release frustration and genuine stress. We know very well that stress can lead to physiological effects in the body and even to structural changes perhaps in the internal gland system(31).

Unquestionably, then, the data processing of conditioned learning reactions has built into it efficient controls and censuring regulators that unfailingly inhibit any association of indefinite or even merely irregular coincidences. Here, too, the organism reacts as a hypothetical realist and considers the coincidence only as a possible reality. Its processor again operates with that double possibility that a coincidence could depend either on necessity or, perhaps, on chance. It depends on the balance of confirmed versus failed (e) expectations, the probability (W) of dominant necessity (W_N^e) versus dominant chance (W_Z^e). In principle, it is the same exponential function of the probability for expecting a law, a lawlike necessity (G_N), which weighs up expectation and disappointment in the probability hypothesis itself. The addition to the comparison hypothesis consists simply of the assumption that we must expect coincidences, and therefore, that we can infer from one characteristic to quite a different one; for example, from the audible sound of the bell to the as yet absent food.

The learning of electronic automata

We already have a good idea of the development of such nerve circuits(32), and through the learning of electronic automata, we have reproduced them excellently since the nineteen-fifties. Attention has been diverted from electronic analogue and digital models to the pure calculation of circuit consequences in the computer(33) and the existence of a uniform principle has been convincingly confirmed (Fig. 23).

Of course, calculation of similarities from the association of coincidences of characteristics develops at the instinctual level much more widely and completely than present computer programmes. Indeed, with the repeated recognition of members of the same species and even more with repeated recognition of single individuals, whole patterns of feature coincidences are processed together in series, gaps in agreement are offset against agreements and, as we shall see, even agreements are gradually weighted as to importance(34). In short, mechanism makes abstraction possible, a process of subtracting the inessential, namely what is unstable and variable and fails expectation, from the essential, namely what is stable, and constant and confirms expectation.

The setting equal involved in comparison, therefore, contains the most justified and necessary process of ignoring the untypical or unforeseeable, in order to make the prospect of the typical and foreseeable more reliable; for the accuracy of aim of the necessary advance judgments on things of importance steadily grows as the predictable is sifted out.

Constancy performances

Amongst the most general of these processing operations are constancy performances that control the perception of temperature, brightness and colour, as well as shape and size. "We all", says Konrad Lorenz(35), "understand without more ado when one speaks of the colour of an object, and we take no account at all of the fact that it reflects completely different wavelengths of light depending on the illumination." The reader sees that this page in the book is white even when he reads it in red evening light or under a yellow lamp. "Although", to quote Bernhard Hassenstein(36), "the messages entering the eye do not truly carry this information faithfully, since it differs according to the illumination and therefore is really falsified; nevertheless, our perception solves the problem of obtaining "pure" information without our assistance, even indeed without our noticing this achievement."

Nowadays, we know that the reporting system can use messages needing correction to revise the message itself, so that, in the example quoted, it obtains the average colour over the whole field of vision and divides each single value by this amount (Fig. 24). This, too, is a universal method that must have arisen a long way back in the history of our ancestors(37). The expectation contained in this operation in turn presupposes the actual coincidence of the phenomena in this world.

The abstraction of form

The most complicated calculations by far are those accomplished in the abstraction of form. This likewise operates through the degree of stability of coincidences and sets aside the unstable from the constant in the most complex hierarchical positional relationships of the characteristics; it eliminates the variable and weights the characteristic. One need only visualise how many characteristics, for example in the retinal image of the outlines of one and the same domestic cat, disappear and re-emerge and the astonishing degree to which they vary with attitude, perspective and distance (Fig. 25). Yet one knows that the actual constancy of such a form is so concealed and so difficult to grasp rationally that we are, so far, unable to store the necessary background knowledge in any computer programme. Nevertheless, our preconscious accomplishes this process of supplementing, abstracting and weighing just like that of the ape or the dog. Indeed, we have reason to believe that even fishes abstract form from its variations.

For, to quote Konrad Lorenz(38) again, wherever "the reaction of an organism is taken in by simple dummies, it concerns an appeal to the innate release mechanisms; where they are not deceivable in this way, it concerns the trained recognition of form". It was long held that the abstraction method of individual learning worked quite differently from abstractive learning of levels of coincidence by the molecular memory. However, G. P. Baerends and his co-workers(39) have supported the opposite as more probable, for example in pig-tailed apes. Ethology and Gestalt psychology have shown rather that the abstracting of form, in species up to man, is merely a matter of more precisely assessing ever more complex material, while the process itself remains almost totally unconscious(40). We must presently show that our own peculiar method of processing similarities, of which we can make ourselves aware, operates with the same hypothesis and the same algorithm that we have come to know from the data processing of any living organisms. We shall recognise rather that neither our world view with its expectation of the comparability of things, nor our language, nor even concept formation could have arisen except on the basis of this ratiomorphic comparison hypothesis; and that rational comparison by itself presents insoluble problems, were it not continually taught and successfully directed by its unconscious genetic instructors.

THE ECONOMY OF EXPECTATIONS

We can now describe more precisely this hypothesis of comparison, of supplementing, of considered abstracting and the making equal of the unequal. Now that the procedure moves into the sphere of consciousness, we can make its components, the members of its algorithm, capable of fulfilment by means of concepts from the body of our own experience right down to details. To make the testing of this hypothesis transparent, let us begin with it at once.

The comparison hypothesis

We can now say, "The hypothesis of the comparable contains the expectation that the dissimilar in the perception of things permits being equalised and that similar things, although obviously not the same, would prove to be comparable in many as yet unperceived properties as well: the similar could be expected to offer a foresight of other similarities."

That this hypothesis, too, must represent an abstraction of the basic structure of our actual world is obvious. One needs only to turn it round in order to see that the opposite assumption prevents us from having any sort of orientation in the world and that every further step we take would become a helpless groping in complete confusion.

If, for example, we see a glistening reddish-yellow apple in a fruit basket, what do we expect? Do we not already see with the organ that we can call our "mind's eye" everything that its surface conceals; the juice, the sweetness, the whiteness of its flesh that the teeth penetrate, or that may be scraped or stewed, the core at which we gnaw and the brown, drop-shaped pips which are so hard and smooth that we can flip them quite a distance by skilfully pressing them between index finger and thumb? Does not a bruised spot on its surface cause us to imagine a soft brownish interior, whereas a tiny round borehole suggests a worm-like inhabitant? In short, the surface of the apple makes us anticipate all the apple characteristics with which we are familiar.

On the contrary, what if we were to expect arbitrary properties inside, say, a bat, or ball-point pens, or thunderstorms, traffic regulations, a volcano or the metal-workers' trade union? Our civilisation which thinks in comparisons would have us put in the madhouse; and, left alone, we should succumb. Indeed, even if we leave open only what would be expected if we approached the apple with mistrust, with asbestos gloves or with the demeanour of an animal tamer; indeed, if we seriously doubted only one single, well-recognised apple characteristic, it would be odd; we should risk undermining our reputation and even our chance of existence. To infer from one similarity to further ones is a biological necessity and therefore firmly built in as regards our expectation about the things of this world.

Inference from analogy

This kind of inference is known as "from analogy", and in drawing it we are inclined to ridicule it, as even our children show(41), for being naive and of little value because we become aware of the errors that must result from lack of experience. Certainly it is naive to expect the sweetness of the apple in a tiny ball, or the bouncing properties of the ball in an apple. However, lack of experience forms the scenario of all cognitive processes; and it is not only the expert of yesterday who laughs at the layman, and the expert of today at the one of yesterday, but, we hope, everyone will smile tomorrow at his own naivety of today(42).

Indeed, this principle of inferring from perceived coincidences of characteristics to those not observed is as universally necessary and established, as inferring from the special to the general. Now we encounter the Hume-Kant-Popper problem of inductive inference from another angle; the solution will be the same, namely that inductive logic is as implausible as probability heuristics is successful.

Similarity fields

All expectations and comparisons occur in imagined fields of similarities, and therefore always relate to a definite group of structures or functions. Expectations as to objects or processes always involve three things: first, that their characteristics would coincide, then that the field would be closed and finally that the distribution of characteristics would allow a common boundary to be expected and recognised and that the perceived characteristics would always be like those to be expected. These three kinds of expectation, taken together, will allow us to determine contents and limits of the similarity field. For neither content without a boundary nor a field without content seems meaningful to us.

Thus, a garage pump attendant who attends to a motor car will expect a radiator cap for the cooling water, and the cook who carves a goose expects to find the liver in a particular place and, with foresight, the most highly specialised features: but precisely with cars or birds only, since, with a motor bicycle or a lobster, the expectations would be very different, not to mention a bicycle or a mushroom dish.

The processing of coincidences

This again corresponds exactly to what we already know; to the level of expectation (G) of the predominance of necessity (G_N), determined from the ratio of confirmed to disappointed ($G_N{}^e$) prediction. For example, if the expectation that a fresh apple surface coincides with a fruity interior, probably by necessity (W_N), turns out correctly let us say, 100 times without contradiction, then the contrary assumption that it might be chance would likewise be proved wrong 100 times. Even if, very generously, we concede every time the chance of 1/2 to the probability of chance then if the probability were (W_Z), it could continue — in this series with a chance of only $(1/2)^{100}$, which equals $(1/1.3) \times 10^{-30}$. The expectation of necessity thus is $G_N = W_N/(W_N + W_Z) = 0.5/(0.5 + (1/1.3) \times 10^{-30})$ which is as good as certainty. For the remaining uncertainty stands at 30 noughts beyond the decimal point; it is merely one part in a quintillion(43).

The succession of confirmed foresight

This example shows the import of repetition, of successively comparable events, the succession of confirmed foresight. More clearly still: if, during a walk through the woods, we observe a dry branch lying across the path, we scarcely notice it and, as we say, we shall soon have forgotten it. However, if, as we walk along, a branch recurs time and again in a corresponding position, then our recollection of it will re-emerge. We find ourselves attentive and wondering whether any more such coincidences will occur. The belief in chance, therefore, will disappear and the suspicion of some purpose, the conjecture that someone here wanted to give a sign, will take its place. Thereby we make a prognosis containing the expectation that we shall find similar signs at comparable distances apart. If this prognosis is not confirmed, then we reject the putative connection and shall forget it. If the prognosis is confirmed at every turn of the path, then we shall soon be convinced that there is a signalling wanderer; after many confirmations, our attention will advance further to the question of the cause behind these signs.

Events occurring only once do not offer us any prospect of something new; because foresight depends on confirmed expectations and this, in turn, on suitable repetitions of the comparable events. Only with known coincidences, for example, the tinkling of a bell when we open the door of a shop, do we take note of it without repeated testing. Repetition is indispensable on the way to any discovery; and the number of essential repetitions depends on the clarity of the coincidence. The classifier studying some new species of beetle can be sure after examining only a few individuals(44) that the little pit, say, on the central part of the head is no accident of birth but is part of the necessary equipment of the species. On the other hand, "The student of behaviour, like the medical man, knows", to quote Konrad Lorenz(45), "that the coincidence of a syndrome of disease symptoms can be taken as consistent only if the observation has been made very often, in some cases literally thousands of times."

The expectation of coincidences is here a mechanism of the conscious reason that can be judged in retrospect but not guided in advance; it is a proto-rational process of the ratiomorphic apparatus. Time and again, we observe that a chance coincidence forces us almost against our will to expect a necessary connection. This may happen even when a window shutter is swinging in an air current in time with a striking church clock(46), where no rational connection could be imagined.

This prejudgment, which makes us expect a necessary and predictable connection wherever coincidences occur, is again far superior to the neutral and unpredisposed processing involved in life; indeed, to such an extent that, firmly rooted in the ratiomorphic programme, it even controls conscious reason. We know this from our attitude to chance. Besides, many experiments show that test individuals undertaking to find a pattern in the flashes or knocks from a random generator usually think that they have succeeded; indeed, when told of their mistake, they go on trying to convince the experimenter that he is wrong in assuming chance(47).

The adjacency of confirmed foresight

The simultaneity of confirmed foresight is processed correspondingly. We experience side-by-side coincidences as differentiation, complexity or as the abundance of characteristics of an object or event. Just as the world grants us a multitude of repeated observations of its objects, it also bestows on us the astonishing abundance of their characteristics. There is no doubt that the principle of preconscious processing rests on these two facts. For just as without repetition we could obtain no foresight of recurrent regularities, so without this simultaneity we could secure no foresight about the composition of objects in the world.

Here, the abundance of characteristics has two sorts of functions or consequences. With the increase in its contents, it determines the growth of the levels of our expectation (G_N), which we experience regarding the identity of the object, and, as something different, regarding the identity of its individual characteristics; and these details are determined no less by what the object is as a whole. Thus, I can recognise the lay-out of a street that I have not seen for a long time, by its details, and these because of their connection in the lay-out. The certainty of recognition (G_N) increases with the number of possible predictions or by the decrease in the remaining alternatives.

We can illustrate this as follows: far below in the valley, something is moving. The number of possibilities is great, but the possible predictions of what it is are very small. On approaching, the details become clearer: a human, a man, a tourist with a cap, our friend H. The number of alternatives drops stepwise to zero, the number of predictions concerning coincidental features increases considerably. It is precisely in this sense that science, too, approaches its objects. The experience gained of constant coincidences of characters in comparative anatomy, histology, cytology and ultrastructural research leaves no doubt as to whether an object belongs to mammals, to nasal mucosa, the ciliate epithelial cells or even to the root systems of cilia. Whilst the alternatives dwindle, the possible forecasts, especially in biostructures, grow to astronomic proportions. Indeed, substructures of a single hair have revealed evidence of over a million single separate characteristics of man(48). Of those, a hundred thousand separate characteristics may be recognised and named individually, which, like a window in a house, or a brick in a large town, may be repeated a dozen or up to a million million million times in the organism(49).

The processing of data again takes place preconsciously according to the number of coinciding separate characteristics, the prediction of which is confirmed by the experience of repeated observations. As with successive coincidences, so with simultaneous ones, the probability level of regular expectation (G_N) is very high with even 10 coincidences, and a practical certainty with 100.

Simultaneous times successive confirmations

Finally, in the overall processing, the characteristics and the repeatability of observation are multiplied, simultaneous and successive coincidence, so that with features that have only ten discernible individual sub-characteristics and repeated observations in only ten related species — with gapless confirmation —

certainty of identity is achieved(50). For the last 200 years, morphological characteristics of identical origin have been called homologous(51) and the "natural classification system" of organisms (Fig. 26) has been based on hundreds of thousands of such revealed homologies. Nevertheless, it is only today that the process, which for long went on pre-consciously, has become a conscious one in its probability processing. Indeed, this has led us into the pitfalls of reason, where we have misunderstood our ratiomorphic instructor(52).

Moreover, form or structure do not consist only of repeatedly observable amounts of coinciding substructures. Rather, these sub-structures additionally show highly special but no less predictable positional relationships and hence interrelated arrangements. These are not only regular side-by-side arrangements in one, two and three dimensions, as occurs repeatedly in our civilisation in the arrangements, perhaps of curbstones, roof tiles, or heaps of bricks, but also a regular arrangement of one inside another, just as drawers occur only in chests, chests in rooms, and rooms in houses, while houses make up streets, and streets towns (Fig. 27). In this, the drawer is itself "only a form of wood", says Carl Friedrich von Weizsäcker(53), "but wood, too, is a form". Substance, for example, cannot be substance of a substance, "but form can be a form of a form".

The hierarchy of structures

The world shows a hierarchical arrangement of its structures and this has the important consequence that certain sub-structures can be expected only in certain super-structures, and these, in turn, must be structures of a further super-structure. Thus, we find that a corner tooth can only be a sub-structure of a jaw, in the skull of the supporting apparatus in a mammal (Fig. 27); and, in turn, the tooth itself can be a super-structure containing crown, root, enamel and pulp; its bone cells contain as supporting substance calcium molecules, atoms and the latter, nucleus and electrons. For experience confirms time and again that a tooth without a jaw, a jaw without a skull, or a skull without an organism can no more function or be realised than a tooth without a root, pulp, enamel or matter. We have still to observe the astonishing extent to which our ideas — and not only scientific ones — are reproduced in these hierarchical structures. Hence, the foreknowledge that we can possess of form or structure contains not only the coincidences of substructures but also complex coincidences of their arrangement(54).

Just as the single structures contain complicated laws about position, so likewise do the similarity fields that they make up. Only within the less complex and their narrowest fields do the structures composing them seem to show no differences; for example the ions of some elements and molecules. However, the similarity field of the elements, the periodic system, is so unequivocally structured according to atomic weights, electron shells and reaction properties, that each member occupies a unique position. This, moreover, gives the similarity fields their structure; and these are differentiated not only by the increasingly complex similarities of the molecules, biomolecules, species and individuals, but differentiate further with size and with time(55).

The hierarchy of similarity fields

Fields of similarities, too, prove to be arranged hierarchically. Together they form a super-structure. This begins to be indicated by the groups of elements, it unfolds with the families of molecules(56) and leads, within similarity fields, to the natural classification system of organisms, from species to genera, families, orders and phyla up to kingdoms into a harmonious gradation of millions of units. Here, with increasing knowledge, every field acquires its own unique position and provides fresh content for prediction. This consists, as perhaps with the perception of the characteristics of a mammal, in assuming all the characteristics of the super-categories of quadrupeds, vertebrates, chordates, animals, Metazoa, and of being constantly able to expect a series of relevant sub-categories according to families, genera and species.

These internal and external structural and positional regularities of the fields offer a further, indispensable content of possible experience. For just as the boundaries of the fields become known through the discontinuities in the changing features of their objects, so do the fields themselves become known not only by differing objects but by the continuity of the changes in their characteristics (Fig. 28). As the sharpness of a boundary depends on the number of coinciding discontinuities of the characteristic, the number of continuously changing features, namely the coincidence of the continuities, determines the unity of the field. In each case this requires detailed exposition(57).

Here, we need say only that this phenomenon of transitions does not signify a limitation but is a further source of the predictable in this world. Frequently it is only knowledge of the transitions that permits us to recognise the connections(58). Knowledge of hydrogen and uranium, of violets and cocoa palms, or of sea-squirts and humming birds alone would never have led to the discovery of the similarity fields of elements, flowering plants or chordate animals. No wonder that we rediscover not only the hierarchical structure of objects, but also that of similarities in the system of our concept formation.

The processing of gaps and contradictions

What if gaps or even contradictions occur in the characteristics of a field? Once again, we can see how much we count on an ordered, harmonious world. For these can only appear in contradiction of an expectation, and hence of a hypothesis about this world's objects. Besides, it now becomes particularly clear how the learning process is made up of expectation and experience, which through reinforcement and failure issue in new expectation and experience.

It is less aggravating when experience shows that our expectation in a field of similarities contains mistakes. As we may remember, we found ourselves quite ready to revise our assumption that all swans must be white. We conform to this learning process in that either we revise the position of the dark swans or the concept of the similarity field of swans. We do not do this arbitrarily, but follow the majority of the coincidences. As happened in the case of the swans, we follow the excess of coinciding discontinuities in characteristics, say, those of beak, head, neck and foot, which belong to the concept of swans, and revise our expectation that we now associate with their colouring.

On the contrary, the experience of contradictions could be aggravating. The discovery of a star-shaped star, or a tree made of bone, or a culture without any communication would cause the respective world view of physics, biology or social sciences to collapse. The discovery of a single anti-coincidence, for example, the presence of a genuine mammalian hair in a species of fish, would alone cause the system of zoology to totter(59); just as Galileo's discovery of the moons of Jupiter, because it was incompatible with Ptolemaic celestial mechanics, brought the geocentric world view into disarray(60).

All these coincidences of boundaries that are involved in the processing or at least the discontinuities of characteristics, the possible gaps and contradictions, are again processed according to their relative frequencies. The hypothesis of probability always precedes that of comparison. We shall not here expound a detailed "biology of concept formation"(61). Here we merely present the algorithm, the solution method which, like a general theory of comparison, forms the biological basis of our conscious understanding of the world.

A universal cycle of rules for abstraction

We find that a cycle of rules is at the basis of the algorithm of comparison, and corresponds entirely to what we already know from the biological discovery of apparent truth. It depends on the already known interaction between expectation and experience, but it goes further, from the processing of single coincidences to that of whole systems of characteristics. Here again the constant and the foreseeable are noted, arranged and assessed above the unstable and uncertain.

This stratum too, the abstraction principle of biological knowledge gaining, extends from the learning of genomes via the individual to the learning of the group. What was hitherto known as abstraction, constant performance and generalisation, we now recognise as the basis of concept formation; as the heuristic guide to the formulation of the hypothetical expectations that we call definitions and laws(62).

Even with birds, and mammals, the abstraction of individual learning, to quote Otto Koehler, assumes the form of "nameless thinking". Today we are quite well informed about the psychogenetic stages of nameless concept formation. As we shall see (cf. Fig. 39), it reaches a simply astonishing extent, especially in primates. Here we shall only refer to the convincing summary by Bernhard Rensch, and to the fact that quite comparable stages in the development of concept formation in children began to be revealed, through the studies of Jean Piaget(63).

The transition to our own forms of abstracting is completely smooth, although we do not wish to minimise the difference as regards reflecting reason becoming dominant; no more than Huxley, Lorenz or Rensch have done(64). We simply want to separate the unreflected from reflected abstraction, or ratiomorphic from rational behaviour, as Brunswik distinguished it, and so find out the possible errors that each of them makes.

Thus considered, it turns out that what we can indeed reflect by way of abstraction is still largely processed unreflectingly, if not exclusively so. We lift the product into consciousness only to let it sink back again, after use, into the unconscious. This belongs to the domain of the psychology of thinking, as developed only in this century under the leadership of Oswald Külpe together with Karl Bühler and others; following Duncker, and more recently through Klix, Lüer and Dörner, it enables us to formulate the heuristic strategies of thought processes(65). The upshot is this: "Contrary to popular opinion, thought itself is not equipped with a particularly high level of consciousness; rather it achieves its results without the intermediate stages being clearly and actually experienced. Finally, there is often the 'aha experience', as Karl Bühler termed it, of a more or less sudden and sometimes quite unsuspected flash of understanding." Investigators like Peter Hofstätter have indeed shown "that test subjects occasionally use a concept correctly for quite a long time but without being able to formulate it"(66). This method not only allows but demands that we count on a direct continuation of the biological abstraction process.

It should not be overlooked that our human thinking receives some guidance from the words and grammar of our language. "But it would be quite wrong to assume", says Lorenz, "that those linguistic processes were the prerequisite of all thought divorced from action. The converse viewpoint that simple clear manipulation in the imagination forms an indispensable basis for every verbal language is far more justified." According to Chomsky, what is universal in our grammar also requires a biological explanation, an inborn origin. "We do not go at all wrong if we ascribe this development to 'natural selection'." Therefore, to turn to Vollmer again, the guidance is mutual(67), and language and thought jointly require a biological explanation.

A biological theory of comparison

It is clear that the process of abstraction, of the imagined making equal or comparing, functions adequately in the domain of unreflective reason; at least enough for us to have found our way about by comparing in the past. However, our reflecting reason, if not unable, is at any rate unprepared to experience this process. With the philosopher, Christian von Ehrenfels, who noticed those preconscious achievements by our reason, the assumption of "Gestalt qualities" arose in psychology, and along with it the still somewhat vague "Gestalt psychology"(68). It arrived at about a dozen accepted rules for the perception of form, like the summation or the "transposability" rule. For example, we perhaps recognise a melody in spite of its being transposed into a quite different key, as indeed every form that we recognise is more than just the sum of the individual characteristics. The stagnation of which Gestalt psychology is accused depends, according to us, on the way it asks its questions. It seeks present-day instead of historical-phylogenetic reasons for its rules of invariance and weighting(69). To the biologist, the life-preserving reasonableness of this control seems to be thoroughly comprehensible.

Evidently, simultaneity and successiveness of the world's regularities are not only processed differently by the ratiomorphic apparatus, but the former enters our consciousness as an experience of form and the latter, intuitively, as a causal experience. Both experiences, according to Kant, are again a priori features for our reflecting, individual reason, namely quality and relation(70).

In understanding Gestalt quality, we must assume that with every perception, however fragmentary, the whole apparently comparable background experience is mobilised; and that the characteristics perceived are considered according to the constancy sought in them and weighed, modified and supplemented within the framework of possible similarity fields. There is constant surveillance by the innate teachers which urge us to count on a consistent and redundant nature of closed comparable forms; and one that is always prepared to re-arrange all prognoses in terms of expanding experience so as to expose them at once as new hypotheses.

Experimentally, we know this change of hypotheses from the concept-forming tasks of classification; for example, by the work of Hovland and Weiss. "If the assumption" of the test subject "is confirmed by the test, that is support for the hypothesis. It is retained. If the report is negative (the classification was wrong)", so Klix sums up "a correction is made or the hypothesis is changed"(71). The strategy of successful classification processes can be simulated in computers. For this we refer to Hunt, Dörner, Klix and Goede. The successful measuring algorithms, through feedback, lead to the modification of the characteristics. And "this change in emphasis", so Friedhart Klix, "reflects the process of subjective formation of invariants through the action of stimuli"(72). A pervasive biological algorithm, therefore, must exist.

We believe that the process itself is independent of the sharpness, extent and level of abstraction of the concept being framed, and merely carried further in the "higher" concepts. As we have already shown for the development of degrees of certainty, the heuristics of comparison likewise contains sliding gradients of conceptual contents according to the degree of certainty, sharpness and the size of field for which they are to contain the prognosis. There is nothing here that corresponds to the step system of Rudolf Carnap(73).

The cycle of expectation and correction

For the second time the algorithm of the biological acquisition of knowledge proves to be universal and closed into a cycle of expectation and experience. What it prepares in advance in the discovery of the apparently true, it can continue in the discovery of what is like in a system of graduated inequality. One limb of the cycle contains the inductive processes of heuristics, the other the deductive processes of logic. The one contains the forecast of the general from particular cases, and the other the controls of membership of the cases from the expected generality. Both advance, circling around the time axis, like a spiral towards the optimisation of possible foresight (Fig. 29).

Likewise, there is a continuum of the products of knowledge which the cycle of the environment extracts by what we call comparison and abstraction, except only that the names of these extracts change. At first these products of perception are called "adapted structures", then successful connections and programmes, imprintings, associations, ideas, concepts, definitions and laws, in the dynamic process of scientific theories.

Again, biology has the largest share in this process. It has amassed two million species, plus five hundred thousand higher units times all their differentiating characteristics. As experience teaches us, it first of all discovered these units by association, then developed ideas by way of trials, turned these into concepts of classes, then defined their contents and limits, and expects that these definitions, the mammal, for example, would be confirmed heuristically as a prediction of the regularity of all mammals in the testing of all species, even those still to be discovered.

Here again we fully agree with Erhard Oeser's cycle of the dynamics of theories (Fig. 29). From the phylogenetic history we can establish why the process of empirical science "is always cyclic in kind"(74), theory of science establishes its composition. It is astonishing that this algorithm of comparison, a prerequisite of all empirical sciences, should be revealed so recently although, as scientific progress shows, it certainly must have functioned in the past.

We attribute this late explanation to the peculiarity of our reflecting reason, which expects something like absolute certainty, at least about some object or other in its thought or perception, so that it might establish and spread this truth with conviction. It mistrusts the vagueness of stochastics, probability and approximations. It mistrusts feedback circuits, the system conditions of causality, and the reaction of effects on their own causes. Our reflecting reason wants to find the spider's threads of a chain of certainties in a multidimensional network of relationships. Science, since the shock by David Hume, has retreated from truth finding to truth transfer(75). Whilst logic transmits truths which are rarely from this world, the empirical sciences obtain truths from this world but not always from that logic. We have mistrusted the process of heuristics, suppressed and forgotten it, and are now astonished that neither induction nor scientific progress can be established.

However, the acquisition of knowledge has made progress because the heuristic principle is so indispensable to the living that it has long since become fixed genetically in the preconscious. It is only when reflecting reason saw itself perplexed that it began to disown its instructors.

The biology of induction

Let us then look once more at this algorithm of comparison, and examine heuristic construction and logical controls from two different aspects.

If we consider the cycle of rules according to the time axis, the succession of processing, then the regulator differentiates into a double loop of expectation and experience. The backward lying part of the loop with respect to time, which encloses the connection between perception and interpretation, we live as experience, and the forward-reaching part as expectation. Together, they form a regression which can be followed back into the depths of the history of living reactions (Fig. 30). The course is such that the sum total of expectations, like that of experiences, always come together with each new content of perception. In this, the processing operation, the comparison of expectation with the new experience, always meets an alternative. If the expectation is confirmed then the new expectation is reinforced, and the experience is enlarged in that specific area. On the other hand, if the expectation fails, then the subsequent expectation is weakened and the growth in experience is at first non-specific and therefore remains to be organised only in expectations of another kind.

On the contrary, if the cycle of rules is considered without the time factor, namely the simultaneous processing of characteristics, then the control differentiates hierarchically between the objects of the general and the special. It regulates between objects and their characteristics between higher and lower concepts, as we say; between super-sets and sub-sets, or super- and sub-systems. Fig. 31 presents the required overview, for, as will be confirmed, the structure of our language is not very suitable to make these hierarchical modes of processing clear. Indeed, that the special is as much an instance of the general, as the general is the law of its instances (example A) will be reasonably evident. However, to see that the special likewise defines the general of the still more special, as well as the special of the still more general (B); or conversely, that the general is defined at once as the special of the still more general and as the general of the still more special (C), we must admit, requires some concentration.

It is the more amazing that our preconscious processing of data operates in this hierarchy so confidently. According to its success, this processing engenders confidence or mistrust as regards the experiences mediated(76).

Type and metamorphosis

To our consciousness, it seems mostly self-evident that we define similarity fields from their representatives, but select the representatives, in turn, according to their similarity fields; though none of these categories is available, we can recognise the type that unites them. Goethe was the first to see the depth of this type theorem; a quotation from his writings heads this chapter. The cause of this recognisable classification, he called esoteric. This has misled the Neo-Platonists to go for the type, but the exact sciences to reject it as idealistic(77). As we shall see, both are quite wrong.

Biological systematics have here established the most amazing performances, in defining the type of each order from the families, and the type of each class from their orders(78); a natural system of hundreds of thousands of types ranging from genera to kingdoms(79). It was only after the event that people realised they did not know how they had done it. Unfortunately they came to think that, since the method was not known, this process could not have any(80). The millipede, when asked how he could walk with so many legs, could not explain it, and afterwards could no longer walk. So runs the fable(81). Meanwhile, morphology has lost its confidence through this and, indeed, has been threatened with total exclusion from the sciences.

This is all the more paradoxical because it was precisely morphology that had penetrated most deeply into the understanding of comparison. From Goethe, who recognised the biological type, to Adolf Remane, who formulated the first rules of comparison, it alone has pointed the way. Now it turns out that the criteria that Remane set up to define similarity of essence and relation are fully confirmed by the biological algorithm of comparison. Remane's main and auxiliary criteria of homology correspond to simultaneous and successive information-gaining in the growth of certainty(82). The homology theorem is the forerunner of our now general theorem of comparison which is its second solution. We should not have found our way without his preparatory work. The theorem rests in the apparent self-evidence of our innate instructor.

We find it equally obvious that in each case we understand the purpose of structures from the superimposed structure system, but its functions from its content and so from its sub-systems. Thus we may recognise vertebral bones from their position in the vertebral column, but the vertebral column from the special features of its sub-structures, namely, the vertebrae(83).

This is determined hierarchically, not because we force nature into our fortuitous hierarchical thought patterns, as people supposed, but because the "order of the living" is itself hierarchically structured and selection had to force on our ratiomorphic apparatus the most adequate system for processing its patterns(84). Chapters 4 and 5 will elaborate on this.

The hierarchy of reason

So it is that a hierarchical order dominates the whole system of class concepts from everyday speech right up to the formulation of scientific theory. Every one of these terms, as soon as we can form it, has its meaning exclusively within a whole series of higher concepts, while deriving its content equally exclusively from all those lower concepts that it embraces. Thus the concept of apple loses its sense if we remove it from the higher concepts of fruit, reproductive organs, plants, growth, organisms. Thus, all form, even of speaking and reading, as Lenneberg has shown (Fig. 32), is developed hierarchically and analysed hierarchically(85). With this method of solution, as we have said following Carl Friedrich von Weizsäcker, "form can be the form of a form". Similarly, a content can be the content of a content and a meaning the meaning of a meaning, and so on(86).

In the third century A.D., Porphyry had discovered the need for this hierarchy. However, the science of thought has as yet gathered little from this

Arbor Porphyriana, this "tree of concepts"(87). The doctrine of concept formation has long presented the view that concepts become ever poorer as they become wider or more abstract(88). This holds only with the tacit assumption that, at any given time, we mentally add the definition of all the higher concepts(89). If so, the concept of apple indeed contains more than that of an organism. If, conversely, we add all definitions of sub-concepts, as biology does too, then concepts become richer as they grow wider(90). Actually, we can ignore neither the higher nor the lower concepts, neither the law nor its instances, neither the sense nor the content of a thing. The contents of a concept as such depend much more on how many features it has and on how sharply it can be defined. This in turn is determined by the level of coincidence of its characteristics within the concept and the degree of discontinuity at its boundaries; indeed, in proportion to how uniform the features are. This indication of a deeper connection within the biology of concept formation(91) is all we can mention here.

What is relevant here is simply that all hierarchical series of concepts remain open-ended; whichever series one follows upwards always ends with concepts such as time, substance, causality, that we already know from Kant's a priori of reason. Whichever series is followed downwards as far as may be, always ends with concepts like point, one or identity, which make up the axioms of our thinking(92).

The series of concepts acquire certainties not, as has been assumed, from their first or last grounds, but from themselves, in full agreement with Oeser's dynamic of theories. In central cases, predictions that can be made from these series, as in biology too, attain probabilities bordering on certainty(93).

The economy of expectations

In these instructions for perceiving form and forming concepts, the process of self-organisation of living things continues; and it is always optimising processes that are selected to increase foresight, probability of success and a steady improvement in the balance between effort and result. We can thus think of them as an economy of expectations somewhat as Ernst Mach did. They are the mirror image of a nature selected on the same dimensions of probability for states and results from hierarchical compartmentalised qualities. Optimised concepts and imagined forms are qualities such as these. This holds for the concept and form of the benzene ring, just as for those of haemoglobin, of amoebae, of primitive man or Gothic architecture.

In the natural sciences, we usually seek for firm reasons and sharp boundaries and think we find both of them in the quantification of qualities; one can, but need not find them there. We can reduce the quality "dog" to the quantities of its organs, tissues, cells, ultra-structures, biomolecules, atoms and to the mean angles and distances of their quanta. However, we perceive much more keenly when we say that we have seen an ordinary dog, the black wolfhound called "Rover", at one year old. If we measure the pyramid of Cheops with a micrometer, this will tell us all and nothing alike(94). However, if we set aside the wide gaps torn by the ravages of time and compare the main features of the structure with available knowledge of pyramids in the Old Kingdom, then the one we shall grasp best is that of Cheops. "The educated man does not push accuracy any further than the nature of the thing requires," said Aristotle.

In this economy of expectations, the structure of open systems repeats itself, the result of self-directed self-optimising abstraction. Our consciousness must first discover its algorithms by stages, in order to achieve the same optimal certainties in representing this world of self-guiding systems.

SENSE AND NONSENSE OF STRUCTURE-EXPECTATION

Finally, we may ask, what would be gained by all this; for it is a useful test. What more would now be known, if we knew what we think we know here? We have seen that a comparison hypothesis is used in the data processing of the living which contains the assumption that similar perceptions let us expect other similarities of perception; and that the dissimilar would again be comparable in a suitable wider sense, and all comparisons will provide a pattern of hierarchical compartments.

The indispensable structuring of the complex

Such a set of expectations would be highly improbable if there were not a corresponding set of states in the actual world. For it can only have been incorporated in the data processing of living organisms by its success in maintaining species. Indeed, this set of states must be reckoned among the indispensable structures of the complex in the actual world. As we admit, knowledge that these structures are real is what makes us confident of expecting the algorithms of the comparison hypothesis to be themselves a product of selection from real states.

In this connection, we may recall that the currently widespread evolutionary doctrines of Neo-Darwinism or of synthetic theory(95) do not yet contain inferences of these states. As far as I know, my system theory of evolution is the first to deduce the necessity for establishing quite distinct patterns for the "order of living organisms", somehow as a consequence of the "strategy of genesis"(96). These are the classificatory patterns of the norm, interdependence, hierarchy and tradition. Here hierarchy, is a special form of interdependence, a mutual dependence of states, therefore, which are moreover encapsulated in each other. These same patterns of interdependence and of tradition are represented in the algorithms of the comparison hypothesis. The assumption that similar perceptions permit the expectation of further similarities reflects the need for interdependence. The expectation that all complex objects, in the widest sense, will correspond to a hierarchical pattern of similarities, is analogous to the encapsulated interdependencies, to the natural order of the hierarchy.

The indispensability expectations of structure

The prejudgment of the comparison hypothesis, the structuring in advance of perception, this advance interpretation or formation that the hypothesis contains, corresponds to an anticipation of the expected natural order. We expect structure to be indispensable because selection has been successful: to expect what will happen is essential for maintaining life. This must simply be so, just as in critical situations right prejudgment must be superior to indecision or wrong prejudgment.

Paradoxical as it must seem to decode our innate teacher, the preconditions of our reason, with the help of that reason, it is equally paradoxical that the sense at the base of advance judgments of the perception of form becomes clear to us only when these prejudgments prove to be false. The first explanations of the biological purpose of optical illusions was given by Erich von Holst(97). We can illustrate this with a simple example.

Thus, it is extremely useful not to assume that, strangely, a distant lion is only the size of an ant; and in view of this vital correction to perspective it is pointless to devote any discussion to it. No wonder, therefore, that our preconscious processing in perspective illusions, shown in Fig. 33, does not countenance being taught by reason. By measuring, we can convince ourselves that the figures are drawn equal in size and, accordingly, the further ones seem much bigger. Likewise for the correction that unconsciously aims at the apparent observation of three mutually orthogonal space dimensions (Fig. 34). It is no different with the corrective completion of figures. As we may recall, it is useful to complete the tail of a lion, at once and without argument, as a part of a whole lion. Again, it is no wonder if snares in completing a picture (as in Fig. 35) are alternately adjusted into a vase or into a double profile. It is natural enough that we are taken in by artificial flowers, or, during carnival, by slices of eggs on a sandwich which proves to be made of rubber, or that we are given the "creeps" in a waxworks, or that a whole cinema full of rational people can alternately be made to laugh or to cry by a constellation of silver grains on a plastic strip(98).

However, it is not only the imitation of nature's classification patterns that forces the ratiomorphic processing to quite specific patterns. Success of thinking in norms, like success of hierarchical abstraction, itself depends in turn on that increase in the chance of success, in the speed of adaptation or comprehension, in prejudgments of storage and rejection, which these successes offer quite universally. This again makes out the economy principle of the living, the condition for selecting what will obtain great advantege for life and survival, with the least effort possible.

The success of hierarchical abstraction

Even in the lottery of vocational guidance, he must win who best succeeds in dividing the professions into equal halves in hierarchic progression. Even with the inclusion of only two thousand professions he has already a hundredfold advantage over one who advises professions singly(99). No wonder, then, that the hierarchy in the system of our concepts is once more furthered when by now we find ourselves predisposed to copy the hierarchy of the system of organisms without our having to know how this happens; that even linguistic expression and the understanding of words and writing is hierarchically built up and analysed alike. The sounds are interpreted from syllables, the syllables from words, and the meaning of words from sentences and even the sentence is stored because it is only from the context in which it stands that we discover how to understand it(100). In the sequel it is still less a matter for wonder that all the products of man, his knowledge, his tools, his institutions, even his scientific theories and all his associations are structured hierarchically(101).

A world of insight, communication and trans-personal knowledge has been acquired through the algorithms of hierarchical comparative expectations and their cycle of rules, and stored in a litre and a half of nerve mass. All this belongs to the meaning of structural expectation; what is experienced, therefore, explains why it is this algorithm that is built into us.

The solution of some puzzles of reason

Beyond this, however, the insight into the comparison hypothesis engaged in structural expectation, gives rise to the solution of some puzzles of reason. The biology of the "hypothesis of the apparently true" has solved some of these already. The "hypothesis of the comparable", which rests on it, develops the solutions further.

First of all, the biology of structural expectation contains the second solution to the problem of reality. The dispute as to whether the world appears to us only as we think it is, because it cannot appear different from what we imagine — or whether it appears to us as it is, because it cannot be thought different from what it is, has been solved. Since the thought that seems real to us is a product of selection, it cannot be more real than the world that has selected it. Moreover, a cycle of rules links expectation with experience for stepwise optimisation, in order that we expect as real more and more of what can be experienced, and experience as real more and more of what can be expected. Thus, the dispute about the priority of reason and experience is resolved. Both revolve as a common spiral that is as long as the cognitive process of the living is old: three thousand million years.

Moreover, the comparison hypothesis contains the second solution to the induction problem of Hume, Kant and Popper. Inference from special to general is indeed not conclusive. However, not only is there a quantitative change in expectational probability with experience, but the structure and quality of expectations are biological processes in the real world: that world's patterns are applied to the very apparatus of cognition that it develops.

Furthermore, the comparison hypothesis solves Kant's a priori of quantity and quality, of subsistence and inherence(102), namely, the problem of what could be the basis of the expectation of constancy and change, type and metamorphosis, since it cannot be derived as a prerequisite for each individual gain of experience, or simply from individual experience. Again our answer is that this expectation springs from the experience of a chain of generations(103). The classificatory pattern of interdependence and hierarchy, which all nature must contain for the realisation of its complex structures, has been firmly incorporated by selection into our world view in the form of algorithms for the comparison hypotheses. These contain the expectation of quantities and the hierarchically abstractable qualities of change and constancy.

The same is expressed for the special problems of our foresight of biological order, when we say: in every affinity group, we expect to recognise the type as well as its metamorphosis. We ascribe to Nature "a rule by which we presuppose that she will proceed", said Goethe, "...and a metamorphosis, which perpetually changes the parts named in the type". This is the view from which we set out in this chapter. We therefore rightly expect to be able to separate the constant characteristics constituting the type (the biologist calls them homologies) from all changes. This is the content of the second solution of the homology theorem, the backbone of all research into relatedness. It proves to be a prerequisite for our knowledge gaining, like the hypotheses of the innate teacher of our thinking and like the a priori of pure reason (the third solution, chapter 4).

All this belongs to the life-preserving sense of our innate expectation as regards a structured world, comparable in itself. As with every advance judgment, nonsense at once follows the sense of the prejudgment.

The nonsense of prejudgment

To begin with the harmless nonsense, even our language can be articulated only in comparisons that are at first superficial. For the landscape contains sand-banks, arms, feet and comb-like ridges of mountains, without anyone sitting, or grasping or running or combing. Then again, there are Adam's apples and eyeballs and, in the language of comparative anatomy, arms of starfish, feet of snails, book-lungs of spiders, and so on, all of which present misleading analogies(104). How else should a data receiver like ours convey to the imagination of a fellow creature even the strangest organism or the newest invention, except by comparison and therefore by equating the unequal? We call an unusual happening indescribable only in order to describe it more colourfully in the language of equating comparisons.

In addition, there is the innate tendency to ascribe structure even to the unstructured. We can think of constellations (Fig. 6) which, once seen, never lose their clarity again even though it is clear that nothing real corresponds to them in the spatial arrangements of the stars. In the same way, mountains look dark, some meadows smile and in the woods at twilight shapes begin to abound; a cheerful world arises, permeated with significant structures; we regret it, when it escapes us.

The boundaries of the selection domain

The real nonsense of prejudgment, however, begins only where we overstep the boundaries of the selection domain and leave the regions for which the innate teachers were selected. On the way to man, we have often overstepped these boundaries. Here begins the domain of good and evil delusions.

Among the good delusions we would include, for example, the innate appreciation that space and time are independent dimensions, time having a linear dimension and space three orthogonal axes. The theory of relativity teaches us about these deceptions. Curved space, however, as in the space-time continuum, cannot be imagined, or only in three-dimensional analogies that are inadequate. Our whole body is built according to Euclidean geometry; eyes, brain, nerve conduction and circuitry (Fig. 36) (105). For us earthworms, selected in a cosmic microregion, the error is quite good enough(106).

Delusions that depend on misunderstandings of the innate teacher are much worse. At one extreme, the misunderstanding contains the belief that there cannot be any innate teachers, because the ratiomorphic processing does not take place consciously. The consequences are extreme empiricism, phenetism and nominalism(107). They maintain that the experienced world consists simply of individual experience, and this contains merely simple images or even only their names. Their delusion lies in taking expectation, theory, as well as abstraction and synthesis as deceptive. From these delusions arise the simplification of the objects of research and a dehumanised science; reductionism, behaviourism, social and cultural Darwinism(108). As a result we are deceived by the sham scientific foundation of an extreme materialism, an inhuman world view.

At the other extreme, the misunderstanding lies in taking as particularly real the product of preconscious processing since it represents the most direct content of experience, indeed thinking it more real than the external world and finally the only reality. This, respectively, is represented by rationalism, idealism and solipsism(109). Whilst we are ruining our own environment, philosophers can continue to argue over its very existence, as Karl Popper has put it(110). In the heavens of Platonic ideas, no other judge remains that could decide between their incompatibilities. Our philosophies turn against each other and the delusion once more is that they are scientifically founded.

The breakdown of the selection domain

The serious evil of these delusions, however, begins with the breakdown of the selection area; where the argumentation, the regulative cycle of the cognitive process ends. That is what causes the trouble: wherever the demands of truth, power and therefore justice link up with half-truths and incompatibilities. There, unconsciously and consciously, begins evil delusion, seduction and manipulation.

Our successful societies have proved themselves specially disposed towards the cultivation of evil delusions. With various technical aids, they can overrun the old controls in man(111), by mass politics and the media they can impose the most intolerable unitary world views, by the conscious, or possibly unconscious use of publicity and marketing, in popular enlightenment and propaganda, which again involves the fact that the preconscious innate teacher cannot be taught by reason. With our expectation of a world that is rationally prestructured for us we accept only too willingly the imprintings on so-called progress, the status symbols, the will to consume and the exponential growth of the golden (steel) cow. Half-truths and incompatibilities in correspondingly reduced selection areas of ideologies arrogate to themselves the status of a scientific substitute for religion. Again all of them, the stupid and the wise, are collectively responsible for collective nonsense. For since legislation has extended the protection of the individual to the borders of a country, the selection of blanket bombing, as is well known, applies without distinction to the whole group. All this must be reckoned part of the nonsense of prejudgment wherever the ability for comparison, fixed in us, is subject to the controls of the self-regulating cognitive process.

Just as with truth and falsehood, we now find like and unlike as the universal antagonists of this scenario, in which the same antagonism of equating the unequal leads us back and forth over the stage, act by act, from the evolution of organisms right up to that of our social systems.

We, the supernumerary actors on this cosmic stage, must admit that we still do not know whether our great ideals of freedom and equality might not contain every creature's freedom to be unequal.

CHAPTER 4
THE HYPOTHESIS OF THE ORIGINAL CAUSE

"It appears, then, that this idea of a necessary connection among events arises from a number of similar instances which occur, of the constant conjunction of these events."
David Hume

"Although the axioms of theory are laid down by man, the success of such an enterprise nevertheless presupposes a high level of order of the objective world."
Albert Einstein(1)

However far the evidence extends into our history, there seems always to have been certainty about one thing: that in all we observe there was always something or other, or rather someone or other, who bore the blame. Early cosmogony explains separation into heaven and earth without mincing words, as the emasculation of Uranus; Chronos, the embittered son, with a stroke of his sharpened sickle, simply separated the embracing pair; and blame for the anger of Chronos rests with the hate of Uranus towards his children(2).

Crime and Punishment

therefore form the third antagonists in the drama of how our judgment developed. They do indeed emerge from the wings late; only after the pairs "truth and lie" as well as "like and unlike" have developed their part. However, these are noticeably up-staged as soon as it transpires round which figures the true drama now unfolds.

Moreover, cause, the Greek aitia, originally meant blame(!) and according to Anaximander's original interpretation, it entails its effect as crime does punishment(3). Some blame could always be found. If a god was offended, a human sacrifice could appease him. If a harvest was spoiled, a witch could be burned for it. If blame on earth was uncertain, then it rested in the stars. Wallenstein's "superstition about the power of the stars made him the last great promoter of Kepler, the founder of modern science"(4). Has not this very science directly confirmed by the old animistic expectation(5) that nothing happens without a cause?

So far, unity prevails. However, as soon as we try to get some idea about causes, whence they came, or merely from which direction, whether they are real, indeed if they exist at all, in all these cases experts differed, and they remain divided to this day.

WHEN THE LIKE WOULD BE THE SAME

You can never step into the same river twice, as the phrase goes; to be sure, where might the water be today into which we stepped yesterday? Nevertheless, the Roman map bearing the inscription Danuvius fluvius teaches us that it was the same Danube two thousand years ago: although neither the drawing nor the letters of its name are the same, its islands even less so, and the water that flowed in it not at all. More precisely, two men may often resemble each other in every detail, but they are clearly not the same. Conversely, a picture of an old man is in no way the same as one of him as an infant, although we know they are the same person. Let us admit that "same" and "like" are neither the same nor equivalent. To be sure, whole series of comparable perceptions have led us to expect that we have the same thing in front of us; at the same time, experience lets us accept radical transformations from an infant to an old man, from a caterpillar to a butterfly or from a village to a town: the wildest change cannot dim our expectation. Thus, we consider individual perceptions as the same if we suspect that they are linked by some principle or continuity.

A second level hypothesis

Such principles or continuities are again only assumptions. They prove to be as indispensable as they are uncertain, being based only on an indirect probability. This shows itself if we take two individuals, say, two herrings, to belong to the same species. No two molecules of them could be the same, yet it seems unnecessary, in fact quite impossible, to trace molecular chains of their seeds, eggs and larvae through the generations and seas as far back as the distant identical two parents; as proof of a continuity from identical origins would require from us. Thus, the same species remains an assumption just as the Nominalists(6) maintain. Indeed, our expectation now rests on a second level hypothesis, namely that the like is not only repeated in the same way, but will also have the same cause.

We always assume this, whether we receive two identical telegrams, or take like matches from a box or like eggs from a nest. Here we assume that the same intention, the same machines or the same hens, respectively, have been at work. We are content with this second level hypothesis, although we were not present either at the despatch of the telegrams, at the cutting of the little pieces of wood or at the hen's laying.

In the end, not even experiment can eliminate the hypothesis. The next throw of a stone will indeed follow a parabola. The cause will be the same gravitational action. However, what gravitons are these? Not the same as in the last throw. Indeed, we do not even know whether gravitons exist. The parabolic paths are to gravitation what instances are to the law, the hypothesis that the same principle is expected behind like expected events. Here, then, hypothesis rests on hypothesis. The hypothesis of the same cause depends on the hypotheses of probability and comparability. We cannot get closer to the cause.

Doubt as to the reality of the cause

It is not surprising that, since David Hume, people have doubted whether our idea of causality actually corresponds to something real in nature. For we can never say: "If or because the sun shines, the stone gets warm", but only: "Whenever the sun shines the stone, too, is warm"(7). Hume argued that a "because" is not a matter of experience but only of expectation. Hence causality is not a real thing, but a requirement of the mind due to habit.

Kant was much occupied with this question(8). For Hume was certainly right: real causality cannot be a product of experience alone. Causality is much more, as we know from Kant; it is a prerequisite for every gain of experience. Again, it is an a priori, a precondition of reason. Nothing can be explained without causality; moreover, reason can find nothing behind causality that should establish it as real.

However, this doubt about the reality of the cause had long been preceded by doubt about its unity. Already in Aristotle we find four separate kinds of cause; they are best illustrated by the example of house building. This requires first of all a driving cause, the efficient cause, namely the application of energy, money or labour. Secondly, material is necessary, the material cause; building material, bricks, cement, beams and so on. Thirdly, it does not happen without a plan determining the shape, the formal cause, which is the plan and elevations that decide the choice and arrangement of the materials; and fourthly, a house is never built without a purpose, a final cause, namely the intention of someone or other(9).

Indeed, we must admit that none of the four causes relating to building a house could ever be left out. Could there ever be a house that had been built without some kind of outlay, materials or without any planned distribution of the materials? Could a house arise without someone or other's intention, however erroneous or misconceived? Clearly not! Even for the dwelling of the beaver or the quiver of a fly larva, none of these four causes is missing(10).

The search for the original cause

However, why should there be just four causes? This confusion by a hypothetical, unreal and fragmented cause leads at once to the splitting of our causal concept. It springs from the search for the origin of a first cause and leads to the contradiction of a discovered original cause.

Have we not always expected that every cause in turn has another, and that causes would arrange themselves like the links of a chain? Must we then not expect a final link, an ultimate cause, from which all others would follow? The original cause was found; but in two incompatible ways.

For one thing, even the early interpreters(11) of Aristotle were agreed that the master may have intended the final cause as the cause of causes. In scholasticism(12), where the interpretation of texts could be taken as an interpretation of the world, this reading was settled. Not only was God's supreme purpose manifest to all, the exemplary causes, which showed that the deepest foundation must be a purpose, in this human world too, it was also obvious that man was first guided by a purpose and in pursuing this he gathered together plans, money and building materials. The causal chain must have had its beginnings in the final cause, in the purpose of the world, in the meaning of the cosmos, in the intentions of the creator. The humanities are rooted in it and have remained close to purposes(13).

Secondly, the modern natural sciences with Kepler, Galileo and Newton, developed quite differently, namely from problems that have nothing to do with aims and intentions. Material and form did not enter into the motion of free fall or that of the heavenly bodies(14). Clearly the question was about forces and impulses that set objects in motion. Of course, the first movement must have been derived from one who is motivated, not in turn moved by anyone else. From the metaphysics of Aristotle, the philosophers knew about the "unmoved mover". It became clear not only that the efficient cause was enough for description, but also, in the sequel, that everything was impelled. The original cause must be force or energy. That seemed enough for the natural sciences.

Doubt about the universality of the cause

The split in this hypothetical, unreal and fragmented causal concept had hardly set into the quarrel amongst the faculties(15) when the so far most modern doubt supervened: whether causality is universal. Again there were two roots. Both emerge from the natural sciences; one from the divisions in biology, the other from quantum physics.

When leading developmental biologists applied the concept of the impelling cause to the phenomenon of self-regulation in embryos, it proved to be an inadequate explanation(16). The regulation could be understood only as a self-differentiation of aims, and this lay beyond what the causal conception in natural science had fixed as an acceptable explanation. The assumption of vital forces that do not obey mere causes, as claimed earlier by vitalism(17), seemed no longer avoidable in the realm of complex living processes. Here too, causality and finality had become opposites, but their general incompatibility will be deferred till Chapter 5. Here we need merely note the resulting uncertainty.

When, in developing quantum theory, Heisenberg formulated the uncertainty principle, a quite different limit to causal events became recognisable in the region of lowest complexity. The tracks of particles proved not to be determined with arbitrary accuracy. It soon became clear that this uncertainty in micro-physics can be extended to the impossibility of certain predictions in the macrodomains of everyday life(18). According to the version of a chain of efficient causes, the first link had become loose and one could ask what remained of the need for the causal concept at all.

Is our causal thinking regulated of itself?

What, then, was the snare? Is reason not rational enough to grasp the concept of causal connection? Or can we leave it as ungrasped when for us an understanding of the world without causality is not possible? Are not science, medicine and technology ever more successful, although we become ever less clear as to what causes are? Is our thinking regulated automatically without our knowing how it happens? Indeed it must be so, both in ordinary life and in the individual sciences. In many modern textbooks causality no longer appears at all. Where, we ask, is the hidden control that ensures our success without our knowing anything for certain about the process? And if there is such a control, we further ask, where are its mistakes? For wherever it is a matter not only of individual areas but of their connection, this control involves us in contradictions and our world view in a vicious circle of incompatibilities.

THE PREJUDGMENT OF REFLEXES AND REFLECTIONS

After all this to-ing and fro-ing between the reefs of reason, it is time to drop anchor more firmly among the hard facts of evolution. Besides, the reader will feel that the solution to those uncertainties is to hand. Otherwise they had not been describable so impartially. However, we must give the solution by stages if we are not to lose sight of the present state of our world view.

Consequently, some solution processes can become complete only as a whole; for example, the problem of final causes can be elucidated only in Chapter 5, although it remains to be shown that people wrongly opposed causality and finality, which quite unjustifiably makes our world view uncertain(19). Here we proceed with causality in the sense of current natural science, namely efficient cause.

The time axis of living organisms

Among the fundamental facts disclosed by investigation of the evolution process is the experience that all living processes follow an axis that we call time. We observed that time need not represent an essential or isolated axis of all events. Time could even turn out to be reversible(20). Physically, time becomes binding only in terms of the entropy law, the second law of thermodynamics. Of its consequences(21), what matters here is that dissipative processes are not reversible. All living processes are dissipative, which means that in all living activity heat is produced and is radiated, that energy, therefore is dissipated in heat, from which alone it can no longer be re-converted. Thus, no living process is reversible and all evolution follows a time axis.

A sequential course for all reactions

In every living event, therefore, there is a sequence, an "if A, then B follows" and state A can never arise again from B. Again, from the egg which will become a hen, there will issue a quite different egg, from the new hen and another cock; even though they may resemble each other as eggs do. Under these elementary conditions it is really trivial to lay down that a sequential course controls all the reactions of the living. The existing structural and operating instructions, coded in the genetic material of organisms, contains the sequential switch "if A, then B", for it has developed from the "if A, then B" of the chemical reaction.

It is no wonder then that all relations to the environment are processed as such sequential relations. Of course, as we already know, the turn-round reaction of the paramecium follows only on collision, piercing by the tick only after the perception of warmth and our own patellar reflex only after sudden increase of tendon tension. Any other processing leads to chaos and, had it been a trial by mutation, would thus have been at once selectively eliminated. More clearly still, in the course of evolution the sense organs for distance anticipate the "if, then" relation. Note how correctly the defence reflex of the suckling is initiated as soon as a collision course of some object is merely suggested. How rationally the hierarchy of instinctive actions arranges the necessary sequence! Thus, the inborn release mechanisms, say in sticklebacks, switch over, first of all, from wandering to the occupation of an area, then they choose between fighting, copulating or nesting, and only if "fight!" had to be chosen do they then choose between pursuit, biting or imposing (Fig. 37)(22). This astonishing "rationality" of the inherited cause-effect programme is a mirror image of causality in the world of the stickleback, built in by trial and selection. The playful learning of, say, the young jackdaw is programmed no differently, even as regards the possible sequences of danger and advantage. It will attack a foreign object, for example, an unknown cushion, first of all as an enemy. If the cushion behaves peaceably, it is investigated as possible food; and when it proves to be indigestible, its liability to disintegration is tested for possible use as material for the nest(23).

The if-then of individual learning

Individual learning, too, starting with conditioned reactions, is therefore already prepared for sequential processing. Thus, in the famous dog exeriments of Pavlov, the unconditioned appetitive behaviour, "if food" — "then the salivary glands flow" is merely side-tracked. As we already know, if the feeding bell were regularly sounded for the dogs at food, then it would simply become: "if the bell sounds, then the saliva flows". It is no different when the rat "discovers" the sequential connection in an experimental arrangement that, for example, "only when the signal lights up, the key is to be pressed" and "if the key is pressed, then food appears". What acts as causal behaviour has been prepared for a long time as a physiological programme.

Starting from this position, only one more step of evolution is needed in order to relate these reactions to the surroundings, which is what the expert calls planning actions, the first stages of insight. As regards this new accomplishment, the step is remarkable, since it requires the development of what we call presentations, representation of space in the central nervous system and, together with the memory contents, the possibility of what we call experimenting with ideas(24). At least, this is achieved by apes. As for the "if-then" relation, little changes except the length of the terms the animal can master, or "survey". We merely alter the name: in place of "if-then" reaction, we say causal behaviour.

Ratiomorphic causal behaviour

We should not yet speak of reason, but this causal behaviour resembles it and is ratiomorphic(25). This is well documented by the numerous observations and experiments on anthropoid apes. In these, sticks are put together as tools, boxes are assembled one on top of the other for reaching food or a branch is provided as a climbing tree to permit surmounting the enclosure (Fig. 38); in the experiment, the chimpanzee, Julia, has learnt to master causal chains of seventeen links (cf. Fig. 50). Her companion, Sarah, who was taught to associate shape symbols with ideas, could even correctly use the very abstract "if-then" (Fig. 39)(26).

We have met the simpler "if-then" programme of a purely physiological kind as the inborn teacher of causal behaviour. Its transfer to performing in the realm of ideas, what we call sensible acting or thinking, must again have been carried through because of its enormous selective advantage. This advantage consists, as Popper and Lorenz alike have recognised(27), in no longer risking one's own skin in the case of actions in the realm of ideas, but rather in merely having to reject the hypothesis. Therefore in cases of gross error, the hypothesis dies vicariously for its owner. That this preserves the species is obvious. The reflection, the raising of the "if-then" programmes into the domain of operations between memory contents, was not aimed at by any organism but was carried through by evolution. There, we now call them the cause-effect programmes.

We are now quite close to the pre- or semi-conscious reflections of man. However, we must first point out two more characteristics of the "if-then" processing. On the one hand, the processing reaches a truly incredible acuteness. The "calculating horse" is an example. Not only does the animal deceive his trainer for years, but the sceptic too, in that it naturally does not calculate, but rather reacts to the slightest expectant gestures from an examiner as soon as the correct number of hoof beats is reached(28). On the other hand, so far we have observed only chain-like assessments of causal connections, and no other kind. About this more later.

The compulsion to reflect causally

The inborn teachers, as Lorenz calls them, which direct the preconscious ratiomorphic reflections of men, are in turn unteachable rationally and therefore not easily observed. However, they show up as a coercion to reflect causally; but of course only if the established expectation is shown to be nonsense rationally. This compulsion consists in expecting coincidences of perceptions to be causally linked, even before the possibility of such a connection was tested. Time and again one may find, for example, that dirty footprints on the floor are linked with one's own steps even though one has come by dry paths. Lorenz describes the experience of having causally associated the swinging of a window shutter with the striking of the clock on a tower in precisely the same rhythm(29). Anyone who remembers such repetitive coincidences will confirm how fully awake one is as soon as the attention is focused on the event and how speedily one turns to a rational testing of the supposed connection.

This behaviour must be a part of the inheritance in our development; otherwise it would not be so incorrigibly ready to advise. Its biological significance for human questions is only too clear. It must indeed have some life-maintaining avantage, as we know from coincidences, at once to expect that a recurring succession is necessarily linked. For in most cases it is; besides, we must surely gain by assuming that it is, while taking note of contrary cases only afterwards, from a safe position as it were. Thus, causes are irremovably entrenched in our expectation; they even enter dreams, albeit often in a quite unreal way.

Teaching us to act

Our idea of causality has developed under the direction of the inborn teachers; what they have taught us is to act ourselves. Under their guidance we learn independently. Even with birds and more so with mammals, the young animal learns incessantly by itself; with man, during his whole youth and with some, throughout their whole life. This autonomous learning from the outset consists much less in deliberately considering processes in nature, but quite on the contrary, in continually interfering with them. They are always activities that guide play and curiosity behaviour, trials, exercises, testings, which are carried out independently. Practising innate methods of movement, the learning of its possible combination, often shade into testing of conditions and the behaviour of objects in this world. Such a prognosis, in turn, furthers the growth of consciousness, though one-sidedly, as we shall see.

Evidently, in such teaching of executive behaviour, experience is first filled with something like the omnipresence of, and then with something like insight into the exclusiveness of, executive causality; this is the expectation that causality can have only one direction and can occur only in chains.

The economy of the elegant solution

For whenever we posit actions, the first cause always seems to be oneself. And only too clearly the causal chains of the action performed appear to run off at once. Expectation always seeks something like the economy of the elegant solution. Whether a young jackdaw tests the nature of a cushion, whether a kitten with its whole palette of possible combinations of instinctive procedures drives a ball of wool across the room, whether an infant grasps a ball time and again, lets it go and then throws it away, whether a child makes sand pies in a tray of sand, destroys them and then remakes them, the course of action always refers to the experience of direct connections. Judgment must be unambiguous, and the most unequivocal connection is certainly the one we write as: if cause A, then effect B. Any other solution involves unnecessary effort; and every living process is assessed according to the ratio of effort to result. The elegant solution will, therefore, be the economical one. It is sought, and therefore found.

That so firmly anchored a principle for solving problems, which is unteachable and therefore still an innate teacher, influences our rational behaviour as well, can hardly surprise us. "Why the easy way when there is a hard one", as the popular gibe has it. However, even in science, when there are two equivalent but competing theories, we like to think that the simpler is correct.

That this is a gross error and the concept of executive causality a gross simplification, can emerge only afterwards. Ernst Mach held that the process of knowledge-gaining conforms to an economy principle(30). If he were referring to biological knowledge-gaining then we think he is right. For wherever prejudgment of the inborn teachers goes beyond the areas for which it was selected, it will lead to nonsense. Still, this belongs to the realm of consciously thinking about the causal nexus, to which we now turn.

The prejudgment of always expecting executive causality has guided evolution wisely; right up to the amazing genesis of man. Need for caution comes only when we consider his amazing capacity for rational knowledge(31).

THE ECONOMY OF CONJECTURES

Life itself, as we have frequently confirmed Konrad Lorenz's insight(32), is a knowledge-gaining process. Not only do the shape of the fish, or the form of the eye, conform to the natural laws of hydrodynamics or optics, respectively; but also the world picture of the inborn teacher, selected for knowledge-gaining, gradually develops in the central nervous systems the most general algorithms for solving cognitive problems. The most general patterns of order in nature are represented in this picture from selection conditions.

As regards regularity in nature that appears to us as the phenomenon of cause and effect, it turned out that the preconscious processor contains the expectation of being able to deal with the predictable sequences of events. To adapt one's own behaviour to empirical advance judgments on the origin and future of states and events turned out to be of life-maintaining economy. It remains to examine how the procedures of conscious reflection, thus guided, must be understood and established.

The expectation of stable sequential events

What, precisely, causes might be we cannot say, as we have seen. We simply do not know whether they are a state of our expectation or whether they correspond to something real in nature. However, what experience teaches us is the daily lesson to be well-advised and always expect causal connections. Such behaviour at first is simply the expectation of stable successive events, namely that we can rely on a recurrent sequence of states of events. This reminds us of what the comparison hypothesis contains, namely the expectation that perception of similarities allows the prediction of further similar features. What was forecast in the comparison hypothesis were the simultaneous or successive coincidences of features, to be expected in an object at a given time or in several objects in successive observations, but in principle simultaneously. Here, however, the time axis is added. For we now indeed expect simultaneous or successive observations, not in order that we might infer coincidences of features, but in order to go on to further successions of features. We rely on being able to expect in one and the same object and in a set of like objects a definite and consistent sequence of states and events.

Cause and effect, ground and consequence

Since we take the time axis to be divided by the present, our expectation is divided as regards judgments too. We expect to possess an advance judgment on states and events that precede the present ones, as well as a prediction of what follows them. In this way we separate the continuum of the imagined sequence of states into what we may term origin and future, causes and effects, or grounds and consequences.

However, our expectations, which we associate with the causal concept, contain something else as well: abstractability. This expectation, too, we know from the comparison hypothesis, and again in its simultaneous form. There, we expected to be able to infer from the special to the general, from the instances in a field, from the similarities of the many to the higher characteristics of the one. Here, we find a similar expectation in the time axis.

If the experience of one, apparently adequate, probability, confirms that the general properties of a set, or a field of similarities, were suitably abstracted from its individual cases, then we expect to be able to make a new judgment in advance. If the generality of the simultaneous characteristics is indeed confirmed, then we consider that we can abstract the general from consequent states or events as well. We commonly call the abstraction of coincidences a description — as we remember from the concepts of class but the abstraction of later states a causal explanation. In Fig. 40, these terms are set out. Some illustrative examples will follow; but, first, some further general features of this process.

Description and explanation

This acceptable, if modest, distinction between two successively formulated expectations we learn from our ratiomorphic world picture. Nevertheless, it has rationally advanced as far as the idea of distinguishing between the so-called descriptive and the causal sciences. It is a distinction which has carried with it a twofold error. One is that it has spread the view that explanations could be found without descriptions. On the contrary, we know that abstraction of consequential states can have a chance of success only after the abstraction of coincidences has proved valid by experience.

Here one may object that the explanation could be perfectly possible even with a single event. For example, that a balloon flies away may be attributed straight away to its being filled with a light gas. Obviously, this succeeds only with the background of prior knowledge, which rests on, at least, an approximate concept of gas-filled balloons. If objects that in terms of their features have been classified as rocks or stuffed birds were then to float off, we should indeed be puzzled, but only as regards epistemology. Psychologically we can at once proceed to explanations, because our ratiomorphic apparatus quickly and reliably sets up unconscious relations of comparison. Clearly this is the path that is usually followed.

Moreover, we often seem to think that prediction of consequent states could be more precise than that of coincidences. The former would simply be descriptive, but the latter, because of the causal enquiry and the susceptibility to experimental testing, would be accessible to the "exact sciences". On the contrary, we know that knowledge of coincidences, as a prerequisite for the knowledge of consequential states, must likewise determine the level of attainable accuracy of consequential inferences. Let it be clearly understood: explanation presupposes description just as much as the hypothesis of cause that of comparison.

The hypothesis of causality

This, the third in the system of hypotheses, contains the expectation that similar states or events will allow us to foretell similar consequent states or events; and that (it again contains the expectation of abstractability) a certain field of similarities, the same set of states or events, allows us to foresee the same definite set of consequent states or events. One sequence lies in the past, another equally definite one in the future; one of them we call ground or cause, the other consequence or effect. In short, therefore, we can say: the hypothesis of cause contains the expectation that similar states or events have the same cause and will have the same effect. This definition of the causal hypothsis must now be tested, applied and established. For in the short form we have merely given a derivation, and one may well ask to what end.

First, we merely confirm David Hume's view that causality need be no more than an expectation. Likewise we confirm that causality, in Kant's sense, represents an a priori, a necessary prerequisite for any individual gain in knowledge, which therefore cannot derive from individual experience alone. What we do gain is its basis. The expectation of causal connections is shown as one of those algorithms proved by selection, which evolution has built into the central nervous system for the purpose of the economical use of data. Like probability and comparison, causality proves to be an a priori of individuals and an a posteriori of their tribe, which has learnt it.

What justifies the a priori in carrying on

From this point of view, we can see why this a priori can go on even in the realm of conscious reflection: because of the same traditional patterns of order(33) in the real world, whose constancy along with selection manages to establish the causal hypothesis in the inborn teachers.

We recall the problem that the present second level hypothesis allows us to expect that two like telegrams, matches or eggs were due to the same intention, machine or hen. We recognise the basis for this conclusion at once if we try to turn it round or simply leave it out.

Let us assume that in springtime we discover a fresh bird's nest in the garden; apparently that of a blackbird. A short climb convinces us: correct! Four blue-green eggs, thickly speckled with red, all alike; and one brief touch (allowed only here) with the back of the finger shows that all four are still warm. What can we make of it now, if we wished to avoid assuming that all four eggs were from the same female blackbird? We should have to postulate some rascally bird fancier who wanted to deceive us. He would have made an imitation blackbird's egg out of gypsum, down to the minutest detail, painted it most deceptively, would have relied on our discovering the nest as well as our climb to examine it and must have counted on our curiosity; for, whilst hidden, waiting for our coming, he would have warmed the spurious egg and, before our arrival, he must have slipped it unseen among the three genuine eggs of the blackbird. At a time when we can travel to the moon and back, all this is admittedly possible, but highly unlikely as a reason for four similar eggs; or at least rather more so than the assumption that all four eggs came from the same blackbird.

It would be equally improbable to assume that someone had wanted to mislead us with regard to the matches that looked so much alike in the match box; and that, to this end, perhaps the director of the match factory personally and secretly, let us say after closing time at the end of the day, cuts a match by hand, dips it, dries it and packs it into the box with his own hands. Thus, what matters is the order of probabilities. Only a joke, if it is good, makes use of its inversion(34). inversion(34).

The structure and complexity of coincidences

Usually the probability ratios of possible hypotheses are processed preconsciously and ratiomorphically. For often we do not enter into the embarrassment of conscious reflection. At best, we reflect on a judgment only after it emerges. Nor can we, in these examples, give any values to the probability ratio. In the telegram example, however, we can show metrically that there is a rational probability of a solution derived from the structure and complexity of the coincidences. According to the measures of information theory(35) the two-fold receipt of the telegram "arrived OK, many greetings, Barbara", for the case in which one assumes the erroneous duplication of the product, the same intention, would have probability 1; if, on the contrary, one had wished to assume that one of the two telegrams did not originate from the same intention but had been due to sheer chance, then this probability would amount to $(1/32)^{35}$, that is, the chance probability of each sign to the power of the number of signs(36); that is, 2.1×10^{-53}, an impossibility for any chance probability on our planet(37).

Clearly it would be absurd not to assume in each of our examples the action of the same bird, the same machine, or the same intention. To leave this judgment unresolved would deprive us, even in everyday life, of any hope that we might orientate ourselves in this world. Therefore, as before, in most cases there are adequate grounds for hypothetically inferring from what is recognised as sufficiently like to the operation of the same cause.

A second level analogical inference

At first this is again no more than a naive analogical inference, indeed at second level. Nevertheless, it is just as essential as the first analogical inference of the comparison hypothesis to which it is linked, being likewise naive but only in the sense of being natural and impartial. However, since now it is to contain a prediction about the cause of similarities, new arguments were found to underestimate it.

In the study of structures, particularly in biology, but also in the disciplines of psychology, sociology and linguistics, it has become the custom to contrast "mere analogies" with alleged insights about essential similarities or true relationships. This partly practical application has been saddled with the false notion that analogies can lead to nothing, or what is worse, only to deceptions. Bare analogy became the warning sign against pseudo-science and the unscientific, although one had to admit that one could not avoid it.

Konrad Lorenz, in his Nobel prize address, began with the solution of this dilemma: "Analogy as a source of knowledge"(38). He maintained that there are no false analogies. That hits the nail on the head. They can exist no more, so we continue, than false similarities. In a similarity only its interpretation can be false, namely the second level hypothesis since it adds explanation to similarity. If one decides:"here comes my friend H" and then immediately adds "incredible, what a deceptive likeness!", the similarity has not changed, only the explanation and that, in turn, reacts on the species, in which one thought one saw this similarity.

Two gradients of interest

Before we explain the reasons for a similarity we must first observe that there are two levels of interest directing our attention and our readiness to elaborate a causal hypothesis. We might call them triviality gradient and helplessness gradient.

The triviality gradient shows that our interest in an interpretation declines with the increase in similarity. When we perceive what is nearly like, we behave as if its common cause were obvious. If two multicoloured lithographs or two young birds resemble each other, as one egg does another, this does not surprise us, although most people would have to admit they had no idea about lithography and, even putting aside all technical knowledge, no knowledge at all as to how genetic information can become a bundle of feathers.

Only when similarities, as between dolphin and bat, between père and Vater, or between Italian and Scandinavian Gothic, become slight is our interest aroused. Where, as between lancelet (Amphioxus) and man, Persian and English, they are somewhat hidden, they become objects of pure science. What we readily do, therefore, is not to determine the mechanism of the common cause, but quite rightly to ask whether one might justify reducing some particular similarity to a common cause.

Again, the helplessness gradient allows our interest to disappear at the other end of the scale, where the prospect of finding a common cause seems to disappear. Faced with the question whether the comparable dumbell shape of a molecule, of a protozoon, of that in a gymnasium or in a galaxy could have the same cause, we might well give up without any regret(39). It is different with the similarity between flowers and the jellyfish mantel (Fig. 41) or between the rhythms of the tides and of menstruation. The problem of understanding similarities between swimming saurians and dolphins, the Maya and Nile pyramids, involves science. Cause thus interests us primarily as what things have in common and as a soluble problem.

Chance analogy

If the common cause of some similarity does not seem clear to us then we first of all behave as if there were none. The product of our interpretation, however, I call chance analogy, largely because one obviously thinks that one can explain such things as the workings of chance. This engages our legitimate interest in seeking to explain our world to ourselves, our ratiomorphic activity enabling us to foresee good or harm; and this leads rationally to the paradoxical view that we have explained by chance what we cannot explain.

Our language, being permeated with chance analogies, may tend to legitimise this view. We speak of Christmas stars and starfish, although the lack of points should exclude real stars.

On the contrary, if we intend a common cause and therefore have to assume the same behind what is like, then the possibility of what we now call a causal explanation splits into two positions. These, surprisingly enough, differ according to the site at which we presume to find the imperfectly known cause; namely inside or outside the objects we are comparing.

Functional analogy

Those similarities that we consider as accountable in terms of causes seeming to produce the sameness from outside I call functional analogies. This fixes the untidy concept of analogy without the need for new words. Typical functional analogies are common in the realm of organisms and all are attributable to the same reaction and adaptation of organisms, of different origin, to the same conditions of their environment. Classical examples include the streamlined form of sharks, sea saurians and dolphins, the eye of the vertebrate and the squid (Fig. 42), the bush shape of hydropolyps and moss animal colonies, and many others. Even more astonishing are the forms of mimicry by which predator fishes (in order to insinuate themselves) and harmless scavenger fishes or harmless insects (for self-protection) strikingly imitate the form of highly defensive species; or even where grasshoppers and butterflies imitate leaves for camouflage, while orchid flowers, in order to attract the pollinator (Fig. 43) are deceptively similar to the female bumble bees(40).

We must grasp that the position of the cause is still in no way obvious from one isolated similarity; how should one, for example, decide whether the striking similarity of two fish, considered in isolation, is due to adaptation or to relationship? Indeed, a decision could result only from the knowledge — or presumed knowledge — of the whole field of similarities, which must be chosen big enough at least to contain all the similarity connections of the partners being compared.

This shows, at once, as for example in the similarity field of vertebrates, that the sub-fields containing the streamlined forms — that is, the groups of cartilaginous and bony fishes, saurians and dolphins — are classified among quite unrelated fishes, reptiles and mammals. They lie dispersed, openly as it were, in the harmonious field of similarities (Fig. 44).

Functionally, analogous similarities have one other thing in common: they are convergent. The further apart from one another their representatives in the common similarity field are placed, the more similar they become with respect to specially considered features. Thus, the highly developed sharks, swimming saurians and dolphins resemble each other more in their streamlined forms than their ancestors, primitive fishes, reptiles and mammals, respectively (Figs 44 and 42).

Functional analogies in the spheres of behaviour, language or culture are explained in the same way. Like the dispersed, convergent similarities, these can only be understood if they are attributed to chance encounters of various internal conditions with the same external condition.

Homology and type

In exactly the same way, we can explore other common causes, of which it must be assumed that the site whence they act is itself present within the system; as Goethe put it, that they are esoteric, or as we say today, system-immanent.

Ever since there has been a science of structural relations, these similarities have been called homology and type, characteristic similarities(41). Many millions of homologous structures have been discovered by comparative anatomists and systematists and, on that basis, two million species have been assembled in the Natural System. They are all attributable to relationships in their hereditary factors. Not only must their characteristics be reproduced with only slight variations, as one might expect, but the degrees of freedom of these features must be restricted according to certain patterns. I have demonstrated this as the cause of the "order in living organisms" and thus as the basis for describing groups of organisms; as a pattern that passes on the conditions of its own history in norms, interdependencies and hierarchies(42).

Of course, with regard to two similar fishes, it is not feasible to decide whether, for example, a similarly formed fin would be due to the same disposition or to the same adaptation. For only the connections of the field of similarities could provide a decision on that.

In contrast to functional analogies, however, closed fields of divergent similarities emerge (Fig. 45). That means that over the whole field of similarities as well as all its sub-fields, one finds closed harmonious changes or metamorphoses, which become clearer the farther apart their representatives in the field. As will be remembered, we judge according to a closed, divergent and hierarchically tiered system of layered structures. What contributes to this pattern is included among the homologous similarities and constitutes the fixations and degrees of freedom of the type that can be determined from the definitions of the groups(43). What cannot be ranged has to be tested for its membership amongst chance analogies and functional ones.

Homologies of behaviour, speech and culture are determined no differently. For how otherwise, to reverse the question, would one explain the harmony of a field of similarities without assuming that the cause of the similarities and their metamorphoses is present in the systems themselves and would be altered harmoniously according to their own regularities?

The third solution of the homology problem

This is a third solution of the problem of homology, this time from the side of its cause. What this cause is like in its details can still remain completely open. The determination of the necessary site of the cause establishes the distinction between homology and functional analogy.

External and internal causes

The differentiation of causes into functional analogies and homologies we have taken from biology, because it is a foundation of biological knowledge and was resolved in that field. The separation of functional analogies and homologies in social and cultural sciences, too, has emerged from biology(44). However, if one is to generalise and include the position of causes in the inorganic sciences, then one must simply apply the conditions that contain the appropriate higher systems against those from the subordinate ones, in place of environment versus heredity. This separation corresponds to that into external and internal causes to which every complex system is subject. It corresponds also to the final and formal causes which act against the direction of material and efficient causes. These we shall examine in Chapter 5.

As to the external conditions for the system, for example, of the individual, what in biology is called selective and adaptive condition, competition and selection, but in society and the community, choice and judgment, inorganic science calls boundary conditions. Whole series of such external systems act causally on their internal systems: the cosmos on its galaxies, the latter on their solar systems, the solar system on its planets, our planet on our biosphere, the latter on its living space, the living space on the species and the species on the individuals; or, the cultures on their groups and these on human beings (we shall come back to this; cf. Fig. 54).

The internal conditions react the other way round; and we speak of the various layers, in terms of quantum laws, the laws of atomic structure, chemical bonds, self-reproduction, heredity, the conditions of metabolism, stimulus conduction and perception, and of the tradition of the contents of consciousness(45). The inorganic sciences have been interested preferentially in internal causes, the social and cultural sciences in external causes. Biology has always been halfway between the two. More about this later.

What a cause really is, we still do not know. However, we can easily resign ourselves to describing the common feature behind which we presume the same cause, as a "general proposition". For such a proposition (its statement is also called a law) permits, for certain properties of a given set or class of objects or processes, prognoses about their states. This alone is of practical importance.

A hierarchy of propositions

The sciences have developed a respectable hierarchy of such general propositions. Thus, each one of these laws contains what we call the explanation of its instances. For example, the law of falling bodies allows prognoses on everything that falls in the terrestrial realm. Its explanation, however, still does not contain a law. We feel it to be explained if, along with others, it becomes an instance of a higher proposition (Fig. 46). Thus, the law of gravitation explains its instances of laws of both terrestrial and celestial mechanics(46).

Just as in the hierarchy of concepts, we never find a fixed point, a beginning or an end. The highest proposition in a given case always remains unexplained and has no higher proposition, while the lowest has no instances and therefore remains unconfirmed. The system of propositions, however, forms a hierarchy of mutually controlling hypotheses and in its midst it contains the greatest near certain probabilities of possible prognoses.

That is what matters. As the hierarchy of propositions grows, so not only does the field of objects in which prognoses becomes possible, but also the degree of certainty and the precision of prediction; whilst the set of what must be known in order to permit the prediction decreases(47).

The economy of conjectures

Granted: our whole system of natural laws may be a system of conjectures, but nevertheless one which permits us to predict more and more in this world with ever greater certainty(48) and with less and less effort(49). This algorithm for the knowledge of nature is based on an economy of conjectures. It is inherited from that economy principle that has proved itself in the prejudgments of preconscious reflections and even from the simplest reflexes and reactions in life, long before selection.

SENSE AND NONSENSE OF CONJECTURES

Let us now leave the details of the object and examine what is relevant and what is false in the hypothesis of causal connections in its given form. We must ask once more how far the inborn method of processing guides us reasonably or leads us into error. We already foresee that this is linked with the real structure of the world and with the area of selection, within which our inborn teachers were established.

In short, the causal hypothesis, as will be remembered, contains the expectation that like things will have the same cause. At first this is no more than a judgment in advance, but as we have seen, it has proved itself in so overwhelmingly many instances that, in principle, it is superior to any other kind of judgment or to abdication from judgment.

A continuity of mutual dependencies

Since this success can at first be understood only as due to adaptation, there must be something in the realm of selection that in the world at large corresponds to this recipe for success. Although we must admit that we cannot know what a cause really is, we must nevertheless expect that the world, whose processes can be predicted by assuming causal connections, will contain a continuity of mutual changes in its conditions. They must be free from chance in time as well.

It is said that Einstein struggled against the thought that God threw dice when deciding whether true physical chance should be recognised as a principle of nature(50). At first it does seem odd that the laws of the world are chosen by chance(51). Until we notice that, if he did not throw dice, he would not have built a world containing free decisions but a deterministic machine with men as automata. "God did indeed gamble!" concluded Manfred Eigen. "Nevertheless, He also obeyed the rules of His game"(52).

The indispensability of causal expectation

These rules, as they arise in cosmic and chemical evolution, in the evolution of organisms, societies and cultures, are handed down inviolable with the objects on which they arise. Physics knows conservation laws(53), biology inheritance, social and cultural sciences tradition. The universality of this constancy or handing of the mutual conditions between objects must be the cause of Kant's a priori of causality; a precondition for natural knowledge. Through the advantages to life which its applications offers, it must have been by selection that it became the indispensable causal expectation. Since objects of this world are hierarchically arranged we can understand that what we extract from its mutual relations in the way of causal and lawlike connections, propositions and explanations, also assumed a hierarchical structure.

Nevertheless, we must here interrupt the flow of admiration for our own excellent adaptation to nature. The insight into the hierarchy of natural laws, as developed by the sciences, by now exceeds what our inborn teacher has prepared in "common sense". Here, for the most part, our conscious reflection has taken over the field. This is different from clarifying to ourselves how we imagine causal connections as such.

It is evident that we "see" causes always and preferentially as acting in a direction and, in complicated cases, in a chain. Of course, cycles of rules have been discovered in nature and have been reproduced in control technology(54). However, we are usually convinced that we can survey the beginning and end of a causal connection. Does not letting an object fall show us where the process begins and ends? Does not the play of billiard balls show how the effects of a cause are unambiguously interlinked? Has not every school experiment suitably demonstrated the beginning and end of chains of effects? Do we not rightly laugh at the well-known psychologist's joke of the experimental rat which boasted to his neighbour rat in the training box, "now you have really got your investigator conditioned: every time you press the knob he throws food in for you". We laugh because we think that only the investigator and not the rat could be the cause of the effects (cf. the solution in Fig. 47). Quite clearly, all causal experience confirms the world picture of the materialistic natural sciences and their achievements confirm the correctness of our views.

Here, we suspect, a new world of errors begins. However, before we describe these, let us summarise what the hypothesis of the original cause has achieved; the wisdom, as it were, and the reasonableness with which it teaches reflective consciousness.

The solution of some puzzles of reason

First of all, the causal hypothesis contains the third solution to the problem of reality. We think of the world in causal connections, not only for that reason, but because we cannot think of it in any other way; but we think of it in terms of cause and effect because nature herself preserves those mutual dependencies which she establishes, along with her objects; and because selection has applied the equivalent of this connection even to our preconscious data processing. Thus the inborn teacher cannot be less real than the thinking which seems so real to us.

Moreover, the hypothesis contains the third solutions to the Hume-Kant-Popper problem of induction and that of the homology problem. As regards induction, probable inference from special instances to its general characteristics is the foundation of probable inference from the special to the general of sequential phenomena as well. There are good grounds for expecting that a similarity field likewise allows inference from like changes in states or events to very probably the same common course, origin and future. That is what we experience as the same in ground and consequence, as the same logical or causal explanation of the like.

As regards the third solution of the homology problem, the inference likewise leads from what is common in the contemporaneous similarities in the organism to the time sequence, to what is common in its sequential states. In this case the cognitive path itself offers two kinds of sites for causes, according to the structure of similarity fields. In the explanatory path, the dispersive-convergent parts of the fields permit inference to meeting the same external causes, the harmonic-divergent similarity fields, on the contrary, to handing on the same internal causes. This is what we separate into functional analogy and homology, which we explain to ourselves from the same adaptation; indeed, from the same disposition.

All this has its roots for a third time in the solution of Kant's a priori; this time in those which Kant calls causality and dependence(55). No doubt, the expectation of a causally interpreted world must be a prerequisite, an a priori, for an individual to acquire knowledge. Likewise it is certainly an empirical product of the chain of generations, an a posteriori of the living as knowledge gaining process. The regression must be as old as the time sequence of biological reactions. All this is a part of the sense of the hypothesis, which here acts wisely.

The nonsense of conjectures

Every innate hypothesis, from the simplest reaction to environmental stimuli up to the elaborate instructions by our preconscious teacher, has limits to its probable correctness. Since they are always judgments in advance they can have high prospects of being suitable only in that range of objects, under whose pressure they were tried by evolutionary mechanisms, tested selectively and firmly incorporated. The further the advance judgments depart from this range the more biased they must become. The nonsense of prejudice always begins at the limits of the selection range, and likewise the nonsense of conjectures. This experience is confirmed for the third time.

The nonsense in the instructions by our innate teacher is rooted once again in the fact that the evolution of reflecting man has long since left behind the limits of what our preconsciously acting ancestors had to recognise and solve. The limits are those within which causality can still be safely regarded as an executive chain-connection and the simplest explanation as the best.

However, we have known for a long time that in the much wider region in which we men have to act and be responsible for our actions, causal connections form a network, in which only adjacent linkages, the threads in a mesh, as it were, can contain a directed linear chain of causes and effects. We were also able to observe that recognising these threads is quite enough for the problem-solving of our animal forebears. What remains to be determined are the deficiencies, the harmless and the evil deceptions, which a henceforth unsuitably narrow causal idea, hard to correct rationally, has prepared for us humans.

The tendency to the simple solution

Reaction of the effect on its cause seems unavailable for processing in our preconscious expectation. This is all the more remarkable when the reverse processing, for example, of the effect of motor reactions on the individual causing them has long been an indispensable principle of data processing. This reafference principle(56) constantly informs us, too, whether, for example, we ourselves shake the bench or are shaken together with it. The assessment of force and counter-force, whether in running, shaking or throwing, has vital significance. In our rational operations, however, it is only too easily overlooked; evidently concealed by our innate tendency towards the simplest solution.

For example, we readily describe the moon as circling round the earth and explain this by reason of its smaller mass; although we know that without mass it could not circle round us, but with mass it must act on a circling earth. Gravitation is a mutual relation as is evident in the tides of the sea.

What, we might ask ourselves, should be the reaction on me if I am the cause of turning this page? This reaction on the cause is actually revealed only by a thought experiment. What confusion would arise if it turned out that suddenly the book had become immovable and my powers and those of all my friends were no longer able to shift it from its position, let alone to close it. Such a physical wonder would be sensational and make the headlines. The reaction we overlooked appears modest. However, it consists in a confirmation of an elementary expectation, namely that the book, given a certain treatment, will close. The continued confirmation of our very trivial expectations is overlooked; although, reinforced and confirmed, it is the basis of all learning, all our orientation and our whole understanding of the world.

Thus, although the reafference(57), in this case the reply of the effect to its cause, as Erhard Oeser in his "dynamics of empirical scientific systems" shows, extends as a requirement for gaining knowledge into the theory of science, it is not provided for in the preconscious of so-called "sound common sense". Even in processing, this backward assessment of the effect on the cause is laborious and discouraging because its aim is so indeterminate; some will indeed admit to having groaned over compound interest calculations. Even our language, by its linear structure, hinders the description of chains of effect that run back into themselves(58). Only the joke profits unabated from this discrepancy between reality and our disposition(59).

The world of deception

Beyond the linear if-then relationships there is a world of deceptions. Even the psychologist's joke about the laboratory rat has misled us. For it is much less funny that the rat thinks itself the cause of the connection. It is more comical that the investigator — as is confirmed by all who laughed at the rat — can even consider himself to be the only cause. For naturally, the ways of behaviour of every training form a cyclic process and every participant is the cause of the behaviour of the others (Fig. 47). Finally, the training also represents the screw of a long regression which, admittedly, could be fully described only with an immense vocabulary of our linear language. In fact, what joins together in training are the natural behaviour cycles of investigator and experimental animal.

However, since it is one of the well-known characteristics of the game to stop when it becomes serious, one can expect that the comical side to our executive causal idea will be strictly limited. It ends, naturally enough, where our own interests become involved; for example, the comical side ends where it meets the conflict concerning the quality of life and environment, which we have ourselves started with our civilisations of success.

Jay Forrester, to whom we owe the first fundamental prognosis of what happens as a result of our growing technocratic world system, says "My basic theme is that the human understanding has not been created to comprehend the behaviour of social systems"(60). Does not the stupid to-ing and fro-ing of world history, the groping of social and economic policy, indeed, the daily confusion in the world news, confirm that our understanding was not created to survey the prevailing complex, non-linear, multi-reinforced cycles(61)? An expert like John Galbraith assures us that explaining the irregularities that occur quickly grows into an almost independent profession, which, with its mixture of reason, prophecy, exorcism and certain elements of magic, at best finds a parallel in primitive religions(62).

It is no wonder that the environmental problem escapes us, and that we ourselves get to know the longer of the linear connections in the biosphere only by means of the ruin that we ourselves have engineered(63). There are experimental researches (Fig. 48) that show us "how men wanted to improve the world and have destroyed it"(64), because, as Dörner found, test subjects think in linear, causal chains and take belated, inadequate account of the network of causes and therefore of the side-effects of their own measures. Perhaps, as Friedrich von Hayek outlined, an individual's reason is in principle barred from understanding the reason of the social system superimposed on him(65).

Here, beyond question, we are in the process of succumbing to the snare of a causal idea, which has long ceased to match the responsibility that we currently arrogate to ourselves in the causal nexus of nature. The methods of instruction and of enlightenment, however, have not yet become aware of this. Perhaps we shall take note of these connections only from the same irreparable damage which we are about to inflict with them.

The evil of deception

These delusions will be bad enough. However, we should not overlook that the evil of deception has a deeper root. It begins wherever deception joins not only with our individual interests but with society's interests in power; wherever the individualist is manipulated by the conformers, the minority by the majority, markets by industry, the masses by the demagogues, and this whole round by ideology. The nearly hypnotic power possessed by the deception of the simplest solution and by that of executive, linear causality, makes the causal idea a particularly suitable hinge for our conflicts. Of the many and in part well-known consequences(66), let us here pursue only one more.

The root of the conflict resides in the scientific world pictures in that field of tension in which they have arisen between myth and reason, metaphysics and the art of experimentation. For "starting from the refining of the mythological world picture, there occurs an ever finer differentiation of the means of thought and hence of the sciences as well"(67). However old this differentiation of our attempts to establish causal connections of the world, the conflict into which our taste for the simple solution has brought us seems just as old. For, as far as we can reconstruct it, even with the development of consciousness(68) it must have become clear that causes come to us from two different sides. No wonder, therefore, that even the oldest philosophers have reflected on the forms of causes(69).

Since our innate teachers suggest the simple executive solution, it is, as we may remember, not surprising that the original cause was no sooner sought than it was also found. The Scholastic of the Middle Ages found it in the final cause and founded the philosophy of idealism; mechanics found it for the Renaissance in the efficient cause and founded the materialism of the natural sciences. The separation of the sciences into those of the spirit and those of matter was completed. With it, asserts Konrad Lorenz, "a dividing wall grew up which inhibited the progress of human knowledge precisely in that direction in which it was most needed"(70).

The double-sidedness of subject and object and the two-sidedness of causes did not become part of the understanding of the world. Rather, two half-explanations began and with their claims to truth, two half-truths became the bases of two incompatible ideologies. Since, as they will tell us, there can only be one truth, Hegel's idealism on the one hand, and on the other — the whole turned upside down — dialectical materialism became the absolutist explanation of the world. Each has been allowed to infiltrate into every doctrine, from elementary school right up into that of reasons of state.

Indeed, all these social utopias have been tolerated, although they have split our world into blocks, although they are about to ruin the world and although, outside the sacred groves of fiction, no further tribunal exists that could have decided which of these incompatible truths might contain the genuine truth.

With the knowledge of our innate teachers, it seems that we may again discover the tribunal which can illuminate the biological background between the incompatibilities of materialism and idealism. We finally turn to the solution of this problem.

CHAPTER 5
THE HYPOTHESIS OF THE PURPOSEFUL

"A philosophy which is not self-consistent cannot be wholly true, but a philosophy which is self-consistent can very well be wholly false."
Bertrand Russell

"To believe in pure nonsense is a privilege of mankind."
Konrad Lorenz(1)

As in a mirror image of the causal hypothesis, we observe once more that, however far the evidence of history reaches, there always seems to have been certainty about one thing: namely, that some kind of purpose or intention always lay buried behind it all. The reader will recall that the Greek word aitia, the oldest known precursor of our causal concept, originally meant blame, which is nearer to intentions than to some physical condition or factor; and that the causes of the oldest cosmologies lay in the aims of purposeful world creators and demiurges.

Sense and nonsense

are thus the fourth and last pair of antagonists in the biological scene that lies at the basis on which our cognitive apparatus has developed. Sense and purpose are first contrasted with the senseless and pointless wherever we feel ourselves involved. It is in these terms that we judge the peculiar features of our bodily structure, our actions, the meaning of our existence,our society and the world in which we live; we call this our power of judgment. The history of this antagonism between sense and nonsense is closely related to that between guilt and punishment; this concerns its biological background as much as the evolution of its becoming conscious and its fate in the rise of culture. Like the causal hypothesis, the sense hypothesis presupposes the prior action of the hypotheses of probability and comparison. Nevertheless, the way we see it, purpose and cause have in turn become opposed. An "anthropology of metaphysics" will, moreover, show us that the consequences of the sense hypothesis must be reckoned among the oldest impulses of the human "spirit". Perhaps what most men understand by the "spirit"(2) of man has the notion of purpose as a precursor in common with emerging consciousness and the sense hypothesis.

Indeed, every notion of man's faith, the simplest as well as the most sublime, starts from the intentions of a purposeful creator. This is opposed only by the materialistic cosmology of the moderns with its views compounded of matter and efficient causes. Transpose the question of causes to before the big bang(3) of modern cosmology and we return at once to notions of faith. Even some agnostics have to admit that in that case they have asked questions about the purpose of the big bang. Even in our daily thinking, we still hold on to the "blame of causes": for example, we ask ourselves what is to "blame"(4) for our car not starting, although what we mean is only the function of the ignition or the carburettor.

The reader may indeed admit the extent of this sense concept, but may ask himself what the tools of biology can do to clarify such philosophical-metaphysical judgments. However, it is precisely the bio-sciences that are continually confronted with the twofold ground of all creatures: efficient causes on the one hand, and purposes on the other. Therefore, an objective assessment of purpose can most readily be expected from biology. In contrast to the rational formulation of purpose, which provoked a problem that involves reality, origin and intelligibility of the concept in an undecidable controversy, we shall continue to establish the hypothesis of purpose from the evolution of organisms as a species-preserving principle.

WHEN THE PURPOSELESS ACQUIRES A PURPOSE

The property of purposes was known to the Ancients: even purposeful structure or form is explicable only from a higher form. The purpose of a lock is to shut a door, that of the door to shut the house, that of the house to serve man.

If the origin of purposes cannot be determined

Similarly, the purposes of a person are derived from his group, those of the group from society, mankind, the cosmos. So there always remains a last undetermined purpose; unless He, God, founded Himself. The origin of all purposes is thus either rationally unproved, or inaccessible(5). It can only, and therefore must, be revealed.

However, if knowledge of nature is the only object of real science, and if man's understanding is not suitable for knowing supernatural finality, then, as John Scotus intimated(6), there would not really be any finality. The first gap has arisen.

If purpose is only an idea

In contrast, Kant clarifies the circumstance that causality alone cannot explain nature adequately. Therefore, it is necessary to seek not only causes but also purposes. For Kant, however, suitability is by no means a detectable component of nature, but rather an idea, something like a control for the human power of judgment(7). A second gap has thus opened long ago. It began with the emergence of natural sciences during the Renaissance and the separation of purpose from cause(8). Cause is now an a priori of experience, purpose an idea of the power of judgment. For German idealism, with Kant, this idea became an endpoint of reflections, the opposite of their starting point; while with Hegel, it became itself, the thing(9).

When the purpose contradicts the causes

This influence could not leave natural science unaffected. Above all, it was the astonishing regulatory capacity of buds and embryos that seemed to resist causal explanations. Consequently Hans Driesch postulated a life force, in the sense of classical entelechy as something that "carries its aim within itself". Thus vitalism arose(10). Philosophers like Bergson extended this into an "élan vital", an objective urge lying at the very foundations of all nature, but beyond understanding(11). In a certain sense, Teilhard de Chardin came to assume a purpose carried within itself for the entire cosmos. Not only was the ultimate purpose not determinable, but the whole world purpose is either a mere idea or inaccessible to the understanding.

The goal-directed cause remains final or teleological, directing the paths and aims of present events, as if from the future; that is, the opposite of causality. Clearly, the main body of natural sciences stands aloof from this discussion. Let the purpose go on behaving as before: they explained the world by efficient causes, from the form of causality valid for them alone.

When the purpose is merely an appeasement

Dialectical materialism is different. It originated with Hegel and his dialectical coherence of the world, but inverted this in a materialist sense, by literally turning Hegel upside down. This did indeed transform final connection into a causal one, but man remained alone with his evident purposes. For if one did not want to expose oneself to the blemish of idealistic error, then purpose had to be limited to the human world and must be put in inverted commas even for the behaviour of animals. Thus Karl Marx in his comparison of master builders and bees(12). Purpose appears only on the basis of the setting up of aims by people, according to V. I. Lenin: "the positing of aims is a manifestation of activity, freedom and the creative character of consciousness"(13).

This is a weak position. For if the existence of purpose depends on the free positing of aims, and this on the freedom of man, then his purpose falls with his freedom. Consequently wherever human freedom is in doubt, his purposes become fictional, mere lies and the dubious purpose of establishing a life bereft of purpose.

As an idea beyond the grasp of the understanding, of an indeterminable end and a doubtful beginning, purpose is for reason the cause of never-ending controversy; but for life it is fundamental.

PREJUDGMENT ABOUT CONDITIONS

To anticipate at once: In the history of nature, what we experience as purpose arose as a cause. Biologically speaking, we could confidently call this a final cause. It stands to efficient cause, and hence to cause in the conventional scientific sense, as the plan of a house to the capital, or the work force to the architect, but it turns out to be an equal member in a system-conditioned, functional causal relation; in the teleonomic and not the teleological sense, as we shall see.

The natural history of mutual conditions

Actually this perspective is as old as the causal problem itself. Even classical interpretation, as we may recall, contrasted efficient cause with final cause. It is only later that our preference for linear executive causality and simple solutions split the connection into two, and was ground away between the millstones of materialism and idealism, respectively asserting that the world can and must be explained only by impulse and purpose. It is only quite recently, after two thousand years of grinding, that the composition of the world in a hierarchical system of totalities could no longer be denied, and with Max Planck, Werner Heisenberg and Carl Friedrich von Weizsäcker, with Paul Weiss, Ludwig von Bertalanffy, with Donald Campbell and Konrad Lorenz(14), that the idea of the final cause has been re-justified. Our own discussion, too, follows on from there. Indeed, impulsive and purposive causes mutually determine each other according to the "strategy of genesis"(15).

Life on our planet begins with the establishment of, indeed owes its existence to, such a mutual relation. As we now know, this was three and a half thousand million years ago, when the surface of the planet's crust had cooled to something below 100°C and collected the primaeval oceans; into them were distilled those interacting energy-rich compounds which were being continually synthetised in the tremendous hydrogen sulphide-methane-water vapour storms(16).

The natural history of purpose

Here begins the natural history of purpose. It has been shown, as Manfred Eigen has convincingly set out, that without the creation of mutual conditions, or mutual promotion of molecules in that warm broth, life could not have arisen, but with those conditions, life had to arise(17). Molecules of nucleic acid were formed — deoxyribonucleic acids — which became the carrier of genetic information. Among many other molecules of different kinds, enzymes were formed, protein molecules that were the carriers of living processes. However, nucleic acid chains, without the protection of enzymes, could not become long enough nor sufficiently rich in information. The chance of enzymes being formed was too small, without carriers of information for constructing them. The selective success of molecules permitted the mutual promotion of nucleic acids and enzymes into the circulation of hypercycles which could then grow away from any other molecular organisations. Only at this stage did they join together into protocells.

Nucleic acids would have remained without purpose, but they possessed the purpose of initiating the building of organisms. Protein molecules would equally have been without purpose, but they acquired the purpose of ensuring the multiplication of their genetic equipment. The cycle between chicken and egg had begun, each on its own without purpose but with purposes for each other. For the egg was made for no other reason than to become a chicken; and for no other purpose did the "free" life of the chicken reach its peak than for securing to new eggs the way into the world of chickens.

The organisation of purposes

Within the sphere of egg and chicken, however, a mighty hierarchy of purposes unfolds between the smallest part and the whole. By selection, the whole organisation of the chicken is directed to correspond with the survival conditions of the chicken world. The environment selects, as a governing condition, which of the overall characteristics of a chicken can be stable within its framework. If the sub-framework "chicken" needs temporarily to fly for its preservation, then wings are promoted. This further sub-framework condition "wing" then determines provision of flight muscle, the muscle fibres and so on, down to the contractile molecules (Fig. 49). Thus the purpose of the myosin molecule is to move the muscle fibre, this in turn the muscle, the flight muscle the wing, and the purpose of the wing is to enable the chicken to fly through the air, the more certainly to fulfil its whole purpose as a chicken.

The entire thousand-fold hierarchy of living structures with its interlinked functions, unmistakable to the biologist, is a similar hierarchy of purposes which are entirely selected to answer to the maintenance conditions of the next higher framework in question. It reveals a community of purpose and organisation, almost running contrary to ordinary idiom. This holds for the chicken, the egg(18) as well as the two together; for cock and hen, for the chicken population and for the maintenance of the entire species. The purpose answers to the fulfilment of a higher function.

The sub-function for a higher function

Our critics might say that, so far, everything related to biological purposes was purposive only in a figurative anthropomorphic sense, since one could not expect a chicken's wing to have the intention of allowing the chicken to fly. It remains to be investigated, therefore, whether there is a connection between the fulfilling of a function on behalf of a higher function, on which the selection-determined differentiation of biological organisation depends, and human goals set by man himself.

To this end we must first consider behaviour. The web of a spider or the comb of a bee colony are products of behaviour. The setting of goals is unmistakable. Who, then, has set their aims? Karl Marx, in his time, maintained that it was not the bees(19). Today we know that it is selection. This operates between two antagonists: between possible mutations as it were "at the level of random creations", which the organisation of the system "bees" allows, on the one hand, and on the other continual offers for optimising life conditions which the higher system "the bees' environment" presents. Obviously, the new task will not be stored under the concept "bees", but in the genetic material. We can assume that this will be so with all instinct-directed autonomous setting of goals; even with the complicated instinct hierarchies of higher organisms.

On the way to such setting of goals within the consciousness of an organism, one thing is retained, namely, the antagonism between the possibilities and requirements between lower and higher systems. However, we must traverse two more levels of differentiation. First of all, the level of individual learning; it begins with conditioned reaction. The reader will remember the food bell of Pavlov's dogs where individual learning depends on association and a new combination of unconditioned reflexes. Moreover, the programmes by which personal experience can be accumulated, as evident particularly from young animals, are themselves arranged purposefully. This amounts to assigning learning goals to individuals, even though remaining under the strict direction of genetic programmes.

Autonomous setting of goals

We reach the level of consciousness where, in higher organisms, it is evident that they begin to seek for solutions in the sphere of the imagination. Konrad Lorenz discovered the biological background of this evolving consciousness, "The backside of the mirror"(20), which therefore allows organisms, even with closed eyes, to reflect on the world of their experiences. Thus, he has described the conditions and stages in the evolution of our consciousness as well(21). A smooth field of transitions becomes evident; in addition to the developing independence of processing under the instruction of the whole stratified system of genetic instructors.

We now find that no later than the primates, goals are already set in the realm of the imagination itself and the solution can likewise be found in it. From the many examples that Bernhard Rensch supplies for planned actions, we may use one to illustrate this(22). The capuchin monkey, "Pablo", fetches one of his toys, the handle of a tennis racquet, grasps it with his curled tail, spontaneously climbs up the cage netting, then puts it between the bench seat and his food dish; then, from a position that is very favourable as regards energy, he tries, with a long lever arm, to detach the dish (Fig. 50). This "self-discovered" game represents typical planning involving the use of a tool(23). The self-discovery of the use of tools is known in nature too. For example, chimpanzees use twigs which they bite into a suitable size, strip of leaves and then use to dig out termites from their nest (Fig. 51); or they may chew leaves into a kind of sponge which they then use to obtain water from small hollows in trees(24).

Between the connection of function to higher functions, on the one hand, and self-fixing of goals on the other, there is more than just one relation. The one contains the instructions for the other; and they represent a mutual principle of evolution. Here, "self" means that, apart from the antagonism between the creative sub-system and its higher system, there is no other participant. "Goal" means the function of a sub-system, of an organ, of a treatment for the next higher system or of an organism that is selected for some environmental possibility. Only the terms for this mechanism of supply are different. We speak of selection, choice or decision, depending on whether the environment excludes what is unsuitable in the genetic material, or the individual what is unsuitable in his environment or in his reflections on this environment.

A power of prejudgment

Such a transition is not merely a sliding along the axis of evolution. These elements of decision-finding merge into each other in every individual. They all contain advance judgments of a particular kind; namely, a prejudgment expecting that a function will always stand the test as a part function of a higher function. Let us call this a power of prejudgment, and let us exemplify its stratification as well as the instructions under which it is formed by the case of organisation in man.

For example, it is beyond question that, regarding our myosin molecules(25), under "instruction" of the higher function, trials on the muscle fibres continued until maximum success was achieved through their parallel arrangement and simultaneous contraction. It is the same with muscles of an extremity which, by the instruction of their common higher function, continue to be changed until at an opposing position an optimum is reached and they are, perhaps differentiated carefully into flexors and extensors. The higher system occurring in the intermediate phase involved can only be composed of the lower systems present. However, the process then appears as if the muscles were aiming at something, indeed, as if they wanted to reach a functional goal.

The patellar reflex(26), in turn, depends on instructions from our whole movement. It automatically stretches the extensor musculature to an extent that the sensors in the knee-cap tendons signal an increase in tension. It contributes to our being able to walk without having to think about it. Again the development of this unconditioned reflex seems as if it had no other goal in view than our upright motion. Actually, it was no less selected under the instruction of this higher function.

The layers of instruction

That whole hierarchies of reflective behaviour patterns can be learnt individually is well known to anyone who, for example, attentively learnt to ride a bicycle. The wish to achieve this here dominates the learning process of our entire static-optic-kinetic processing. Whilst learning, a-large part of our consciousness becomes anxiously overwhelmed by the very varied messages of perceptual mistakes; until, with the goal of riding safely achieved, all sub-functions sink back again into the unconscious layers of the connecting network. It is again clear that this goal can be reached only between the antagonism of our body structure and a bicycle-positive environment. This goal would not be open to us potential cyclists in a tree-top environment, nor, within a cycling club, to a cart-horse. If it were suggested theoretically, in some fit of creativity, selection would destroy the projected course from the start.

What if the goal is Euclidean geometry(27)? A civilisation and its requirements built wholly in accordance with three space axes points the way: and an eye and a brain perceiving and processing in the same three axes, confirmed by three rotations about the same three axes, allow this to be followed smoothly and indeed automatically. The automatism is prepared genetically; considering what is new in the learning situation, it becomes conscious and then sinks back again into routine; or automatically re-emerges when, having turned to the right at right angles four times in a strange town, we find ourselves back again but not at the same place, contrary to expectations.

The case of the "autonomously set goal" of building a house now follows smoothly. Again all the levels of our organisation are involved in the realisation of the goal. Once more in the higher system, in our society, all instructions are prepared for the materials, craftsmen, brokers, solicitors, down to the building regulations, public opinion, credit and standard symbols, and again in the lower system, the architect's brain, decisions regarding the number of occupants, capital, site, situation, style and the purposes of the rooms, have been anticipated long since. Whether the possibility eventually becomes a house or whether it remains a castle in the air, depends largely on fitting together these higher and lower systems. Room for manoeuvre is indeed minimal with respect to what can be realised in the fields of goals offered.

This should not suggest that there is any doubt regarding free will or that any lack of goals is being asserted or even the special role of consciousness is being denied. The small amount of freedom that we experience as self-decision, however, is not peculiar to our own species. It is the creative principle of evolution; only it is in turn called mutation, then association and finally a voluntary decision. It contains the creative freedom of each of its strata.

As to consciousness, this too is seamless, indeed it becomes possible only through all the deeper layers, from preconscious or unconscious reflection, through conditioned and unconditioned reflexes down to the simplest reactions of living organisms. Reflection, the reaction of life to its world, shows steps in evolution and layers in the individual. Both are boundless and follow the second, cognitive principle of evolution.

The goal-setting of evolution

There is no lack of goals anywhere in the organic world; but goals are always determined by the higher system and their feasibility only by the lower system. For species and for individuals, environment determines the goal and the population or the individual may or may not achieve it. Our human environment is determined by our culture, civilisation, politics, ideology and religion. These define almost all goals, but only some of them are reached by some individuals. Even the greatest and the most free among us, with our self-imposed goals, have progressed only a little way beyond the goals of our times(28). Evolution sets the goals with the higher systems in question, and the most successful steps of their lower systems are always small.

Conscious expectation

Let us go back to the advance judgment that guided the tuning of organic organisation. We have established that it contained something like an expectation that every function would prove itself as a part function of a higher function. This takes our layer comparison so far towards consciousness that we must now ask how that expectation becomes conscious. We can again assume that as regards this expectation the mode of conscious processing will follow the primaeval principles on which it builds.

For a fourth time, we must therefore ask how our conscious judgment finding behaves towards our experience or feeling of purpose. For a fourth time we can examine whether this special kind of judgment finding, strange as it is, could be understood from an instruction by the innate teachers. Our own history from primaeval man right up to modern scientific theories offers abundant material.

AN ECONOMY OF THE MIND

There is nothing to indicate that some organism or other would become conscious of itself. Rather, we may remember that it must have been the extraordinary, life-preserving advantages of operating in the imagination, in the centrally-represented spaces that permeate consciousness as soon as the prerequisites for its creation coincided. This revolutionary possibility of pitting experience against experience, now in the space of memory, showed itself well prepared, seeing how it processed experiences in mutual assessment. We found that a probability hypothesis sorted out chance from necessity; that a comparison hypothesis, built on this, separated like from unlike; and finally, we found a causality hypothesis established on the first two, a second order hypothesis leading to the assumption that behind the like there was something that we imagine as a chain of the same causes.

Rationalising the causal direction

This executive causal processing of our innate teacher, operating according to the “if-then” principle, must have been the godfather to our becoming conscious of or experiencing causality. The executive algorithm, selectively programmed into the unconscious central nervous system for causal processing as the most economical pathway to solution, now had to direct problem solutions in a world of events with which unidimensional causality evidently cannot cope. In a multidimensional system of causes, the task of discovering its seemingly single valid dimension had to result in rationalising the causal direction.

Whence do causal chains come and whither do they run? On the one hand, it had to become clear to our awakening consciousness that causal chains begin with our own actions and go on from there; from taking a stone, throwing it, following through its flight and then its crashing down, scaring a flock of birds and ending with a few lost feathers floating down. On the other hand, even early man must have realised that causes begin beyond his grasp, come nearer and terminate at himself; for example, the approach of a storm, a flood, or a stone thrown at him. Must it not be natural, when experiencing one's own intention and its execution and resulting consequences, to suspect an alien intention behind the events to which one now finds oneself subjected? Like the purposes we experience with our own activities, must there not be behind storms, floods, seasons and all becoming and decay, somebody's purpose or intention, as a last cause, indeed a very last one beyond the world?

An anthropology of metaphysics

We recall that what here sounds like an anthropology of metaphysics is amply documented. Not only does the forerunner of our word "cause", the Greek aitia, mean "blame", but the oldest cosmogonies extant begin with the wrath and persecution of a highly purposeful world creator. Still older are the ice-age cave, skull and bear cults. Even though Neanderthal man thought that something further happened after death (Fig. 52) or believed, as people of the Arctic do today, that the bear was a kind of mediator between man and the spirits dominating the world, he was nevertheless prepared to search for its metaphysical causes(29). Anxiety, says Lucretius, was the first mother of the gods. Belief, the delegation of intentions to realms beyond possible experience, was surely the earliest spiritual impulse of consciousness. The universality of religio has a deep and irreplaceable basis.

This interpretation of higher causes was of course reinforced biologically. It must always have been vitally important to recognise that, with all one's subfunctions, one was only a part of a whole series of higher functions. Indeed, this expectation has long been prepared by the organisation of purposes in the organic, as we have seen. It merely interprets them rationally. In either case, life success determines the subordination of the individual to the higher functions of the sex pair and to those of the group; the subordination of the group under the predator-prey system, as rationalised somewhat strangely in the appeasing ceremonies of the fertility and skull cult, and the bear cult respectively. From cave paintings down to mythology, we are shown how the differentiating consciousness imagines it can appease the higher functions of hunting, group welfare, weather and seasons(30).

The hypothesis of purposes

This last in the system of hypotheses contains the expectation that the functions of similar systems may be understood as sub-functions of the same higher system; in other words: that like structures correspond to or satisfy the same purpose. For example, we need recognise the function of a pair of scissors only once, in order that we should have to anticipate essentially the same purpose in roughly similar structures, such as wire cutters, candle snuffers, crab pincers or block shears. We shall have in view that this fills our language with analogies, for example, in leg scissors or scissor-tail. For analogies are mixed depending on form and function. The compulsiveness of such an expectation, similar to "magical thinking", has been investigated where it fails as well(31). We do not go astray in expecting that every joint, every tube would correspond to some function of movement and conduction in a higher system; that the muscle in a leg, the leg in an individual, the individual in society, would fulfil its purpose; and does so even before we think about it.

"We have, therefore, some idea of teleology in nature, in an a priori manner", as Immanuel Kant remarked, "and the possibility a priori of this kind of notion, which is not yet knowledge, rests on the fact that we perceive in ourselves the capability of linkage according to purposes (nexus finalis)"(32). Still, we must not overlook that this a priori of the purpose hypothesis has arisen at a higher level of evolution. If we could conclude that the hypotheses of probability, comparison and cause were learned from the genetic material of our kind, then the hypothesis of purposes, even if under the instructions of the innate executive causal hypothesis, presupposes a presentiment of the "I"; namely, some judgment on the direction from which the cause acted on ourselves. Accordingly, we do not find this a priori developed under Kant's a priori of pure reason but in the critique of judgment(33).

How purposes arise

The decisive question now is: whence do purposes derive? For the final cause has always been "a stranger in the natural sciences"(34) and has remained so until now(35). Thus it becomes the touchstone for the reality of purposes to ask, in the style of scientific materialism, whence have purposes come? To start with, science finds no indication that purposes could have existed before the creations of evolution. Besides, the cosmic, chemical, biological and cultural evolutions exhibit the chronology of a layered structure in which the complexity of the world consistently increases, layer by layer, from quanta, through atoms, molecules, biomolecules and individuals up to societies and cultures (Fig. 53). Only extreme idealism, such as Hegel's, assumed that the more complex not yet in existence could be the final cause for creating its previous layer. How, therefore, could purposes arise before the objects that fulfil them?

The solution is simple and binding: new purposes always arise between the part and the whole. Whatever the structure of organisms has differentiated in the way of living purposes, and whatever has been developed in the course of three thousand million years in the way of suitable cells, tissues, organs, drives and instincts,always arose as a new intermediate layer between the functions of the species and the molecules of its genetic programme. The highest purpose is always the same: the preservation of the species. From it issue the chains of purposive causes, determining form and function in every leg of a crab, in its pincers, its muscles, their fibres and their myosin molecules. We recall the example of the chicken. We know the mechanism that runs through this. Genetic changes continually provide for variation, and selection takes care of the choice of the more successful, the more economical, which finally appears as the wise solution. Causes, however, act from the higher layers and finally from species maintenance, which, like a goal, is retained in all those creatures that so far have survived selection. "Therefore, in teleology, one speaks", scientifically, as Kant had anticipated, "quite rightly of the wisdom, the economy, care and beneficence of nature, without thereby making it into an intelligent being"(36).

The purposive cause as uniting formal causes

If now these final causes remain constant from layer to layer insofar as the common purpose is always and only the success of survival, they differ in the formative conditions which selection lays down. Obviously, the functional and formal conditions that selection prescribes from the survival requirements of the crab's extremities, are different from those which the extremities impose on the pincers, the pincers on the muscle and so on, down to the myosin molecule. The common purposive cause is only the unifier of all the formal causes of the living. Where the formal causes change from layer to layer, the purposive cause penetrates uniformly into the organic. They do indeed share the direction of their action. With both, the cause remains in the conditions of the higher layer and in both it exerts its effect in the appropriate lower layer. This brings us very close to the first solution of an old cognitive problem.

"The world", so R. Eisler commented on Kant, "is so arranged that the powers and laws governing in it lead to a suitable development"(37). And "it is a maxim of reason that even what has the clearest link with purposes has nevertheless arisen in accordance with the order of nature"(38). Kant leaves open only whether the nexus finalis is an objective principle of nature. "We do not know if it is simply a subtle and objective empty" notion(39); "a subjective principle of reason for judgment, which as a regulator (not constitutive) for our human power of judgment is just as valid as if it were an objective principle"(40).

Are formal causes a principle of nature?

However, having discovered the connection between final and formal causes, we must ask whether formal causes are a principle of nature. We find that formal causes, in the whole of nature, contain limiting preconditions which the appropriate higher systems impose on the possible maintenance conditions of the lower systems. The principle is homogeneous except that its name changes with the layers, between which we observe its effects. In the inorganic sciences it is called boundary conditions (Fig. 54), in biology one speaks of selection, competition and breeding selection of individuals and of the adaptation of their organisation; in behaviour, civilisation and culture, of decisions by voting, judgment and reason.

The reality of the principle is not in doubt. For only too clearly do boundary conditions in the cosmos determine the form of its galaxies; these, in turn, of their solar systems; these, their planets; the planets the possible associations within them; the latter, the choice of their atoms, and every atom, the number of exchangeable quanta. In addition, all these formations are differentiated insertions between the totality of the cosmos and its smallest parts, the quanta (Fig. 54). If the layer of groups of living organisms is interpolated between a planet and its available surfaces and atmospheres, then the hierarchy of formal causes is extended quite considerably. For the environment selects the species, whose individuals compete and choose and the formal conditions of the individuals determine the adaptive organisation of the organs, these the tissues, the tissues the cells, the cells their biomolecules, and so on, down to the last vital hydrogen bond of a molecule in the molecular fibril of genetic information. If, between the environment and the species, the packaged layer of society is inserted, then this chooses its groups and the latter the acceptable individuals, and so on. The process continues to the layer differentiation of cultures, voting decisions, reason so-called, according to their own laws.

The objective effect of formal causes, therefore, in whole layers of the real world, consists in limiting what is possible to the most stable sub-systems under the maintenance conditions of its relevant higher system. This is the first solution of the problem and it indicates, as required by idealist philosophy, that causes indeed act from the whole to the component parts; in the opposite sense to what the materialist world philosophy allows. For Kant, this was "a quite different kind of original causality" which to him still seemed as if "an architectonic understanding was the basis of nature"(41).

The limits of purpose

However closely our concepts of purposive and formal causes run parallel and determine each other in the organic world, at its limits we sense the limits of purposes too. These may be readily discovered by following through the layers of the real world what, among formal causes, we feel as a purpose.

The purpose of our intentional actions seems quite evident to us. If we follow the layers outwards and question our own purpose as individuals, then it certainly needs some reflection to determine it. We are part of mankind. If we ask for their purpose, then we become uncertain or we need to borrow from the realms beyond experience. Human beings are unquestionably a sub-system of the biosphere, this, in turn, a sub-system of the planet and beyond that of the solar system. However, our concept of purpose ends at the biosphere or, at any rate, on the planet.

The penetration of purposes

We do indeed follow purposes right down to the last vital molecule and its quanta. The water molecule, which still finds purpose in cooling the skin by its evaporation, has lost any purpose for us as soon as it floats away. The precious water acquires no less a purpose in our understanding when it assumes its life-preserving functions, as when a thirsty man finds it and strives after it. We find a penetration of purpose into the lower layers. They arise when they assume functions for living things and surrender them again as soon as these pass. Both limits are "reasonably" determined by our suppositions and they change with our knowledge and inclinations.

The idea of purpose thus evidently depends on whether we think that some sub-function with a reasonable chance of success were in a position to satisfy the life requirements of its higher system, as we should like to imagine. So it happens that we ourselves may consider the most elementary life function, such as the search for water by a thirsty man, as pointless when we feel he has no hope of finding any. Thus we recognise the purposes of cogwheels, shafts and transmissions, even in a purposeless machine, because the intentions and the success of its constructor are clear.

Purpose as an honorary title

Purpose, therefore, is a respectful title for the penetrating, unchanging formal causes that we think we can compare for the realisation of our own life functions and intentions; like the concept of harmony, which represents an honorary testimony to proportions that we find pleasant. Purpose, like harmony, originates from an admiration for ourselves. "Purposes", says Kant, "have an immediate reference to reason, be it a stranger's or our own. Only, to place them in another's reason, we must base this on our own at least as an analogue; for without this they cannot be imagined at all"(42).

For the derivation of a priori from an a posteriori experience of our innate teacher, purpose is a third-order analogical inference. We inferred from the similar to the structurally like; and further, from the like to the causally identical, and beyond this we infer to the same higher function from like sub-functions.

Of course, like every judgment in advance at the beginning of experience, this analogical inference is as naive but also as vital as its precursors. It certainly promotes our maintenance conditions if, for example, faced with the presumed wrath of a sergeant-major, a bull or a disturbed swarm of wasps, we try at once to escape, without prior musings; we simply let ourselves be guided by the analogy with the consequences that our own fury might have.

However much purposive cause may be an honorary title for some naive analogical inference, moreover of the third level, there exists an area in the real world to which it fairly precisely corresponds; namely to that unity of all formal causes for the area of the living, which is so important for us.

We must now step back from our subject in order to take up a perspective that emerged when we began to discuss causality, but could not then be established.

Materialism and idealism

At the start of the discussion on causes, we met with the unsolved problem of a world with four causes; in addition we met with an attempt at a solution which lay in a desire to find the original cause of all causes. In this attempt one finds that materialism and idealism are incompatible, a contradiction that survives in spite of two thousand years of cultural history; at last now we can remove it.

We can now see that, within the network of functional causal connections with respect to the layered structure of the real world, causes running in counter-directions do really exist. So long as we followed quite generally the "hypothesis of the original cause", we were mainly following the traditional materialist concept of the natural sciences that, since Galileo and Newton, thought the impelling or efficient cause was the only scientifically comprehensible one. As soon as we come to objects of greater complexity, we must distinguish between external and internal causes, which have been shown to run counter to each other. In the "hypothesis of the purposeful" we developed from formal causes the scientific foundation of purposive or final causes, which, although a "stranger" to our natural sciences, have always formed a pillar of support for the humanities and the causal concept for the idealist account of the world.

Now we find the four kinds of cause (Fig. 55) in a symmetrical relationship to each other; efficient and material causes act outwards from the deep layers, final and formal causes from the higher layers downwards through the structure of the real world(43). Whilst the material and formal causes change from layer to layer, the efficient causes as forces penetrate through these unchanged, as do final causes, at any rate as purposes in the organic world.

Indeterminism and determinism

If the four causes thus deduced form a mutually dependent system, then another perennial problem dissolves: that of determinism versus indeterminism. We recall that doubts about the unbroken lawlikeness of the world emerge from the materialist philosophy of natural science; that real chance was discovered in the layers of microphysical processes; and that this indeterminism can extend into the macro-regions of the world. If we bear in mind that the creativity of organic evolution is due to the fact that organisms, through chance mutations of the genetic material, have retained that microphysical chance, then all life would appear to be indeterminate. Since chance is the opposite of purpose, materialists like Jacques Monod felt that man should at long last understand that he can have neither sense nor purpose in this world(44). However, we feel that this cannot be right, we experience purpose.

The interpretation of idealism which must emerge from the highest world purpose of a creator runs in the opposite direction. From this, all evolution becomes the necessary deterministic consequence of a comprehensive intention. Thus idealists, and in the end Teilhard de Chardin too, have deduced it(45). This is again incompatible with a world predetermined from the beginning, without room for the creative, for responsibility and for the presupposition of free will, which did not escape the idealists either(46).

Our solution is a world view in which determinism and lawlikeness arise from chance, in that the necessary creative chance is trapped by fortuitously emerging necessity. "God does not gamble!", Albert Einstein repeatedly maintained, "He wagers", as Manfred Eigen says today, "but He also observes His own rules of the game!"(47); namely those rules which He won in the antagonism between sense and freedom. We cannot pursue this subject here. It is just one of the further consequences of our "biology of knowledge", although it forms a basis for the knowledge of the "strategy of genesis", which I have described in detail elsewhere(48). We must now concentrate on the question:

How purposes acquire sense

and particularly in the scientific sense of true goal formation, of a sense of direction pointing to the future and hence to the problem of the teleonomy of real processes. If we are not careful, this, according to Wolfgang Stegmüller, will cause us to "enter an equally time-honoured and almost impenetrable primaeval forest of philosophy"(49). Indeed, teleological explanation considers a goal as cause; and "what happens in such an explanation is nothing less than an explanation of a present event by reference to future conditions and processes"(50). Actually, causal goals acting in the future are not at all what we have in mind. For what we have discovered in the way of formal and purposive causes is exclusively distinguished by the direction of effect in the layer structure of complexity but not in the time-direction of the causes of impulses and materials. This is crucial: we call this teleonomy.

The co-operative effect of the two, internal and external cause, promotes a particular kind of biological learning processes. The phenomenon of homology and type indicates that not only experience of the environment is incorporated in the genetic material, but also experience from the purposes of the organism's organisation is quite firmly taken up into the programme. The fact that we can define, and repeatedly find confirmed, unambiguous units, such as beetle, fern or mammal, in other words "the classification of living things", is attributable to this learning process. Under that title(51), I have substantiated this process from the specialist point of view. This is too technical to be given here in full. For the question of how a sense of direction comes into the course of evolution, the following account may suffice.

Structural and operating instructions, coded in the genetic material, do not remain independent of each other. On the contrary, selection has to choose those mutants in which chance interactions between such genetic messages arose as are coded for structural components that have like functional connections, on the basis of their much increased speed of adaptation(52). From the success of its products, the genetic material thereby learns a special logic or grammar which, with increasing complexity, can no longer be replaced with increasing functional responsibility and is less and less subject to alteration. The basic features of all affinity groups are fixed once and for all. So we possess

The fourth solution of the homology problem,

henceforth the functional foundation for development of the type and structural plans of organisms. That is why notions of systematics describe real things in nature; why a mammal must always remain a mammal, why the adaptability of the structural plan of the mammal no longer allows, say, a bird or a fish to be formed, although selection has tried to do this through environmental conditions with the bat, or the dolphin, for some hundreds of millions of years.

The fixing of a group of characteristics or the laying down of some systematic unit is no singular event; rather, layer upon layer of groups of characteristics are irrevocably determined. So it is that in the genetic material of man, for example, the possible adaptability is gradually narrowed down to the features of animals, of the chordate, the vertebrate, quadruped, mammal, placental animal, primate(53) down to the genus Homo, and the species Homo sapiens. An entire hierarchy of prescriptions stretches the curve of variability increasingly into the time axis (Fig. 56). The concave curves of systematics arise(54).

Such ancestral curves, the spread of which we see documented in fossils over at least five hundred million years, permit certain prognoses about the future as well. We are forced to assume that they will extend similarly into the future; any other expectation would be unfounded. Linked with this are the phenomena of

Orthogenesis, sense of direction and teleonomy

Orthogenesis describes the fact that pathways of evolution have taken a direction(55). The sense of direction is the first consequence of this; it contains the extrapolation into the future of the direction adopted. It permits prediction of what is possible for the evolving systems in future and which fields of purpose remain open to them. Again, this is a biological consequence of teleonomy. For every evolutionary pathway there is something like a goal, a field of future goals. This ultimately consists in it becoming less and less probable that fixations will be overcome, the higher the fixed system becomes ranked. Thus the boundary of the species is pierced time and again, those of the genus and family rarely, and those of the order and class scarcely ever(56).

It is imperative to understand here that nothing acts teleologically from the future into the present, and that there is no natural law that intends something or would have a goal in view. There are interacting limitations due to material and formal causes, which continue to guide self-developing systems in one direction within the maintenance possibilities or purposes remaining to them. The sense of direction arises with an accumulation of these purposes themselves. It is the product of numerous chances and subsequently of their establishment as necessities. Therefore, the direction which the course follows can be called a chance. In the end, however, every fixed pathway, even if unpremeditated, necessarily aims at a very definite field of goals. A purpose, being tied to its definition, is singled out from its variable features and thus dignified by what we experience as sense.

For man's future, this means that he will never again be able to escape from the laws of vertebrate life, the mammal, social associations, verbal communications and reflection. He no longer can increase his chance of survival by, say, a rearrangement of his external skeleton, by laying eggs, by a solitary mode of existence or, indeed, by discontinuing communications or thought. His chances of survival are to be found rather in more successful adaptation of his hands, his speech or his reason. It is interesting to determine how very much this remaining field of purpose, this sense of direction, unfolding of dexterity, communication and understanding, agrees with what we feel as our own sense.

The economy of the mind

Summarising then, we conclude that the rationalisation of the hypotheses of sense and purpose developed in our species corresponds to an economy in the management of our mind; this is based on three levels:

First, the purposive cause contains the expectation that most functions of the organic world and therefore most of our activities will be sub-functions of some function of a further higher system. How could I "according to the peculiar state of my cognitive ability"(57) decide otherwise for my dealings, if I did not assume, however vaguely, and indeed under some circumstances falsely, that they had been put there for the use of the next higher life function?

Secondly, the purposive hypothesis recognises a whole hierarchy of purposes. It runs parallel with a graduated system of formal causes that together determine maintenance and survival conditions for the same hierarchy of real systems. The pervasive unity of these conditions we experience in the middle range as the purpose of things. Since we possess no organ for experiencing the extremities of this hierarchy of purposive causes, we are accustomed to the relative certainties at the centre, as in the experience of every causal connection. How could we live if we did not suppose, even if not very definitely and perhaps wrongly, that ultimately all our other life functions would be functions of yet further functions; if we could not suppose that the purpose of our activities can be based on life's purpose, and this in turn being difficult to recognise can be based, if not on the circular inference from the purposes of our activities, at any rate quite generally on the hierarchy of purposes?

Thirdly, one can become aware, by making the layered structure of purposes conscious, that a sense of direction and the goals of evolution arise from this; and that these fields of goals, as yet open to our survival chances, coincide strikingly with what seems to remain to us as man's sense, even if vague and to an increasing degree uncertain. Since it seems that only few men can consider life as making sense, without at times assuming a further sense to their individual existence; and as this sense always becomes subjectively more uncertain through the contradictions of ideologies and their insufferable claims to being right; the insight that evolution has developed our objective sense along with ourselves, may continue to contribute to the economy of our mind.

SENSE AND NONSENSE IN PURPOSES

It would now be highly superficial to suppose that the hypothesis of purpose prepared in us, acted only for good. We have learned too much of the ambivalence of the achievements of evolution to expect this without question. It would not be very responsible to finish here. Moreover, one will already predict where the nonsense of this fourth advance judgment about the world begins; namely, at the limits of what can be tested, at the limits of selection. For obviously, as we know, an advance judgment that must have been developed in ignorance of the world as a whole for the purpose of gaining knowledge in the world can only be correct for that sphere for which it was continually tested. Within the selection area it becomes a wise instructor. Beyond that, it becomes plain nonsense, systematically misleading and deliberately used for deception.

The differentiation of relation

The hypothesis of purpose contains, as will be remembered, judgments in advance to the effect that, behind functions of similar systems, the same particular function would be expected with reference to the same higher system; and that like structures would have the same purpose. Let us therefore first investigate its successes.

In the unconscious, with knowledge gained ratiomorphically, this hypothesis leads to the wisdom of relying on a predictable differentiation of relations. In organisms, there arises an enormous hierarchy of sub-functions and sub-structures of purposes. “This is the notion of purpose in nature . . . like that of the structure of eyes and ears, of which, however”, as Immanuel Kant still had to point out, “as far as experience is concerned, there is no further knowledge than what Epicurus conceded, namely that after nature had formed eyes and ears, we use them to see and hear, but this does not prove that the cause leading to their formation would itself intentionally have formed these structures for the purposes named”(58). However, we now know that nature certainly had no intention, but continually had to select from all chance trials those whose sub-structures proved themselves as functions of a higher function, namely of species preservation. We cannot even say that it had been nature’s purpose to preserve species, but in conformity with selection, only those are retained which, in their turn, as a function of the next higher function, answered to the maintenance conditions present in the environment. Purposes arise precisely with their systems.

Purposes are thus selection products of the life functions of sub-systems within the framework of the maintenance conditions of their appropriate higher systems; they are the pervasive unifiers of a whole hierarchy of formal causes inserted between the part and the whole; for the preservation of the individual, the species, of society and its culture. It is part of the "strategy of genesis"(59), that even the carrier of experience to be passed on, namely the genetic material, learns not only the whole purpose but the entire hierarchy of purposes.

The recognition of our differentiated dependence

The same maintenance conditions for differentiation directs the rationalising of purposes. We then experience it as the recognition of our differentiated dependence on superior formal causes. This recognition is, in turn, of vital importance. Contrary to the interpretation of dialectical materialism, no antithesis is to be found between architect and bee(60). This selection of sub-functions under the instruction of higher functions, whether we call them layered boundary conditions, selection, breeding, judgment or reason, is all the same to the biomolecules, bodily function, genetics, reflectively or rationally learning individual. All recognise the actions of higher causes as a prerequisite for the maintenance of their existence; the effect of a formal and final cause, which does not operate from the future into the present, but as against efficient and material causes, from the whole to its parts.

In recognising purposes, one merely prepares the notion of a many-sided causality as it informs our experience against the inconsistency of our innate hypothesis of an original executive cause. No activities, no life plan, no group or culture could survive which did not recognise itself as a sub-function of a higher function. The reversal of the hypothesis of purpose, by itself, shows its prudent, life-preserving guidance. If we merely assumed that behind the similar, basically different purposes are to be expected, then our success — even our life expectation — would be drastically reduced. Cervantes has pointed this out through Sancho Panza.

We must learn to unite the two sides of causes. "The possibility of such a union of two quite different kinds of causality", wrote Immanuel Kant, "is not grasped by our reason; it lies in the supersensible substrate of nature"(61); so does four-dimensional space or the continuum of space and time. This question has remained open for over two thousand years, since Aristotle and Epicurus. Even for Nicolai Hartmann, causal and final nexus lie too far apart to show directly how they become overlaid by determinations. "What lies in between can structurally be only guessed"(62). It is only the presentiment of many-sided causality that is prepared for our perception, its rational establishment is a matter of science. How it became many-sided we learn from the mechanism of evolution.

The solution of some puzzles of reason

For the fourth time we find the solution of some puzzles of reason. What is central is the insight into the equal justification of the two causal directions and the knowledge that each can itself explain a good deal but not everything. Neither the materialist nor the idealist concept can be defended as the exclusive explanation. Kant, too, has affirmed this(63). However, what for him remained unquestionable as an a priori of pure reason and of the power of judgment, we have established a posteriori, from the cognitive process of evolution.

If materialism and idealism are revealed as half-truths, so their scientific consequences, reductionism and vitalism, likewise prove to be unfounded. In their interpretation the reductionists err in attributing all causes to those of the next lower layers. The vitalists err in postulating that causes must be assumed which operate from the future.

If the effect of formal causes is recognised, then we gain the fourth, the functional solution of the homology and type problem. The type is then based on the genetic learning process that is initiated by the adaptive success of functions of its own products. The sequel is a causal foundation of the natural system of organisms, of orthogenesis, of the sense of direction of evolution and its fields of goals. The teleonomy of evolutionary pathways proves to be the result and not the cause of its sense of direction.

Finally, it follows from the above that our world can be neither destitute of harmony nor full of pre-established harmony. It has post-established harmony. Its development stabilises itself; it is not wholly determinist, nor undetermined, its products are neither predestined nor arbitrary results of chance. They are predisposed to come together in new regularities. Whenever evolution creates something new, the world inevitably contains the chance of creative freedom and it guides this into the trap of accidental necessity; what was predisposed acquires new regularity, to the new sense of direction of its pathways. Sense and freedom are the mutually conditioning antagonists in the strategy of genesis.

The belief in pure nonsense

This chorus of promising solutions, however, should not be allowed to sound without putting before us that ample measure of nonsense that is no less derived from the advance judgement of purposes. The nonsense begins where extrapolations become pure prejudice, and where they depart from the limits of what can be tested. Again it turns out that: “the belief in pure nonsense”, as Konrad Lorenz has said, “is a privilege of mankind”(64). This interesting fact is explained in that even the apparently most senseless modes of animal behaviour are never without some measure of correctness, for they are merely extrapolations of the biological sense that they have learned.

For example, the irrationality of insects buzzing to death on a window pane, or of soil organisms, as mentioned before, that are unable to avoid falling to their death in funnels, may be attributed to the fact that highly life-preserving decisions in the selection area were incorporated in their genetic programmes. The decisions were, respectively, that if the flight path is barred, the greatest safety is in the direction of the light; and that in case of threatened dryness, safety will be found in the deeper soil layers.

Given man's ideas of purpose, things can be quite different. The purposes that he thinks he needs for understanding his subordination in nature usually wander off into a realm of ideas beyond any possibility of his control. This must land him in a domain of demiurges, witches and spirits; and from the idealist arrogance of seeing the purposes of his species in the whole world, arises the foul evil of superstition.

"The speculative interest of reason", says Kant, "makes it necessary to regard all orderly arrangement in the world as if it had sprung from the intentions of a highest reason"(65). It is a "lazy reason", he continued, served by all "the purposes, often made only by ourselves, to make it very convenient for ourselves in investigating causes, namely . . . directly appealing to the inscrutable decree of a highest wisdom"(66).

Collective nonsense and collective blame

Now individual nonsense may not disturb us very much. If our neighbour thinks it advisable, he may process his anxieties with the skulls of slain cave bears, with Poseidon, angry angels or with "big brother" who he is convinced is always watching him(67). It turns out, however, that even in the human sphere nonsense is usually a social product; it is collective nonsense. Whoever wishes to ascribe the individual nonsense of his neighbour to the creative powers of individuals, will have to admit that people are somewhat doubtful about this notion; that finding others of like mind is not alarming, but on the contrary reassuring. This is dictated by the wish to understand and be understood, and initiated from the fundamentals of our world view which tells us that the number of times an expectation is confirmed must have something to do with its correctness.

According to the laws of evolution, the price for security in collective nonsense is, of course, that of collective blame, and hence having partly to pay for the nonsense of the whole group. The controlling regulator that may still act as a correction, in cases of individual nonsense, becomes a group liability with collective nonsense. Someone or other must know what are the true goals and world intentions of mankind and so the group knows what the real truth is; therefore it now becomes highly acceptable either to bury the followers of kings alive(68), to sacrifice men's hearts, to burn witches or even put kings and aristocrats to the guillotine.

Being misled by metaphysics

It may now seem to us, writer and reader alike, because we too have survived gassings and bombings, that even the performances of collective nonsense appear merely as dark phases in the world's history, especially as the history books already try to convince us that the majority of its battles may have been fought perhaps for some good to mankind. Even if this were so, one need only bring to mind the immediate consequences of human nonsense in metaphysical deceptions. Metaphysics, to begin with no more than a technical term(69), through Neoplatonism(70) became a science of what lies beyond experience, the supernatural; knowledge that no one can test. Throughout the two thousand years of our cultural history it has remained a major discipline of philosophy, the science of what one cannot know. Now, along with the metaphysicians, we accept that metaphysics is a necessary activity. However, we have confirmed that this depends on one of the peculiarities of our world view, that we feel impelled to seek for the reasons for existence, reality, the essence of the world and, further still, for the superordinated reasons. We also concede that a colourful world of supernatural purposes can be assumed in this field. At the same time, we observe that, in consequence, we can no longer trust any further tribunal that wants to distinguish between true and false. In cases of dispute, it is a miserable judge.

Deception by ideology

Now even the most colourful world of pluralistic abundance of opinions would be no ground for fear. Rather, ground for fear arises from the claim to truth itself by the most incompatible metaphysical systems and the scientific legitimising which, with their false certainties about what cannot be known, they have presented to ideologists. Ideology, once a fashionable word of the Enlightenment, but for Karl Marx those fixed forms of thought on which a social order is based(71), arises as soon as political demands and finally power demands unite with those "certainties" about some arbitrary purposes of mankind.

Then, simple error changes into evil deception. The hypothesis of purpose, at first a source of errors beyond control, becomes the lever of the demagogic method. Humbug becomes deceit as soon as it is shown that the promotion and when necessary the implementation of humbug becomes a successful political instrument in mobilising passion, if only for producing an image of the enemy for the purposes of ruling(72).

Once again, it becomes evident how deep-rooted is our expectation of general purposes, how unsettled our individual views of a shared sense, how easily led the collective mind and how susceptible to brain-washing are those who deviate. The evolutive control of these contradictions, the selection of the wrong things, which, in the shape of collective nonsense of the group was invariably effected with fire and brimstone, becomes a conflict between world powers for tribal security, a collective responsibility for a collective guilt, suddenly to be borne by all.

In our innate teachers and consequently in what we call our sound, unreflective common sense, we have discovered for the fourth time an indispensable guide to possible gains in experience. We know that common sense instructs us wisely within the sphere of natural controls, but brings us continually into difficulties beyond that. Indeed, all teachers have been established in the environment and for a knowledge of the world to which they optimally correspond. For our peculiar cognitive capacities they remain a prerequisite for every gain in knowledge. To the extent that our brain has enabled us to transcend the simple environment of our ancestors, we should have given it species-preserving controls of its own; these correctives are simply more knowledge and more humanity.

CHAPTER 6
SOLUTIONS AND THEIR CONSEQUENCES

"Evolutionary epistemology accomplishes . . . a truly Copernican revolution."
Gerhard Vollmer

"In this model there is no absolute beginning . . . there are no first facts and no last grounds."
Erhard Oeser(1)

We must now sum up, a somewhat sketchy synopsis for those who want to know in a hurry; but also a survey of the position, results and consequences of this study. What we have done is to set down a theory, obtained by application of the techniques of biology, and it seeks to understand the cognitive process as a procedure embracing all creative learning, from the knowledge gaining of the earliest living structures to our reflecting consciousness. We have looked for an ancestral basis for our reason. The purpose of this investigation was a closer determination of man and the sense and nonsense of his possibilities.

The study was suggested by the insight of epistemology, indicating that human reason cannot be based on itself alone, and the insight of ecology, suggesting that there are consequences of this reason that currently endanger our existence. The study was made possible by the insight of evolutionary theory into the development of patterns of order in the real world and in the continuity of that mechanism, which constantly gains knowledge from that order. This mechanism must, therefore, be our main concern.

How biology teaches reason

What epistemology alone cannot do, biology can. It can provide the observer with a site outside the objects being investigated. It examines (Fig. 57) the gradual development of patterns of order, the formation of learning mechanisms developing within them and the learning results from the information provided by biomolecules and biostructures, as far as ways of behaviour. All this is prerequisite for the emergence of our conscious reason; it is nevertheless an independent external object and therefore may be investigated objectively and comparatively.

In that way the separation of objective and subjective cognitive structures is achieved, the establishment of agreement between patterns of nature and patterns of thought, "the formulation of hypotheses that are testable empirically, dealing with inborn cognitive structures and of hypotheses regarding their phylogenetic development"(2). All this will help to build evolutionary epistemology into a complete theory.

Since this standpoint lies outside the processes which make up our reason, what is rational in them and under what conditions they lead into error may also be objectively determined. Moreover, since one can follow the evolution of these conditions until they become opposed to our conscious reflection, the interesting question of what is unreasonable may be answered as well.

WHAT BIOLOGY CAN SOLVE

A theory must first agree with the experience in its intended sphere of validity; otherwise it would be contradictory. This "agreement" is in our view substantiated in Chapters 2 to 5. They contain typical examples of the objects of experience to which they are intended to relate. A theory without extensive explanation would, in turn, have no value; so we shall summarise what they are able to solve. Finally we require, along with Karl Popper(3), that our theory may well fail to stand up to future experience; otherwise it would not be testable. We therefore classify its objects according to the problems to which it offers solutions.

Let us begin with a brief survey of the problem of knowledge and the structure of the theory.

The trilemma of cognition

Perception, inference and meditation, intuition and revelation should all lead to knowledge. What about the certainty of our knowledge? Can one prove all or at least some of it? The basic postulate, according to which all assertions should be proved, leads into a threefold blind alley, which Hans Albert has suitably termed the "Münchhausen trilemma". To quote Gerhard Vollmer, "one has only the choice between (a) an infinitive regress, in which one goes further and further back in search of grounds, (b) a logical circle whereby one goes back to statements which, for their part, had already occurred as basically essential, and (c) abandoning the method"(4). The regress is infinite, circularity cannot lead to experience, and to give up is to admit that one cannot base reason on itself.

Is there no knowledge that would be certain in itself? For centuries, people were convinced that there must be an Archimedean point of absolute certainty. Descartes, for example, sought it in the evidence of consciousness, Pascal in the theories of geometry. Today we know that all "concepts and basic laws" of any theory "are the free discoveries of the human mind", as Albert Einstein has somewhat pointedly formulated it. Even the laws of mathematics insofar as they "have reference to reality, are . . . not certain, and insofar as they are certain, they do not relate to reality"(5). Therefore, there can be no certainty as such.

Our model corresponds to this insight. That Archimedean point does not exist. What Einstein says of the propositions of mathematics, must apply to those of logic. They could transfer truth only if it were there(6). Is this the end of the blind alley, into which epistemology has stumbled?

Actually, this is no end. On the contrary, epistemology as we see it becomes a section of the biological process of cognition and it has foreseen, even if fragmentarily, the spiral structure of our model. The regress is indeed almost unending. For three and a half thousand million years, structures with an age of 1.2×10^{10} years have been learning(7). The cycle of expectation and experience would be a circle if the expectation were not changed with every experience, and conversely. In the same way, the tracing back of learning structures must be broken off wherever these no longer contain their object; for example, that of the system of consciousness in the nerve, that of stimulus conduction in the transport of substances, and that of inheritance in chemical reactions. The end of the methodical reduction must lie where fulguration has led to new system laws.

The evolution of reason

We consider as rational the choice of those decisions which contribute to maintaining and improving the circumstances of life. To create such a reason needs variation of established decisions and a constant selection of suitable ones. In nature this begins with mutation and selection processes; as Manfred Eigen shows, with the first cycle between molecularly coded structural instructions and the selection of their product if, in turn, this itself promotes the reproduction of its instructions(8). This cycle between trial and judgment may now be followed up to the point where we experience it consciously as a cycle between expectation and experience. It represents an algorithm, hence it is a cyclic process which, through the repetition of certain processes and a limited number of cycles, leads to optimisation in the solution of a problem(9).

The evolution of this algorithm lies in the building of layer upon layer of learning and the functioning of the layer is a prerequisite for formation of the next one. In that way the principle of the algorithm is copied under the instructions of the previous layer and only the means, the functions of its parts, are refined.

Six parts may be followed diagrammatically in these cycles throughout evolution: three on the expectation side and three on the experience side (Fig. 58). All six consistently represent the learning results of the whole cycle from the successes of all the previous layers. Moreover, there are transitions between the layers. These are not included in the diagram of Fig. 58. The seven layers roughly correspond to stages of evolution, as would be attained by the cell, nerve cell, the brain or consciousness.

The experiential contents become more differentiated as the possibilities of the system increase. The mode of experience extends from simple increase or decrease right up to verification or falsification; so, too, for reaction to experience. The same holds for the drives of what we summarily call expectation; from the energetic and simplest physiological conditions of life maintenance to those which we experience as intentions and goals. Furthermore, the aids to decisions are of interest; for, with increasing possibilities of the system, the number of possible decisions becomes boundless. A learning mechanism that cannot do without chance for its creative steps, cannot allow the repertoire of chance to go beyond certain limits(10).

For the winning chance, in the case of equal chances, always corresponds to the reciprocal of the number of possibilities. Therefore genetic coding, the system of gene interactions, the hierarchy of instincts, innate release mechanisms, prejudgments and arrangements right up to so-callsed self-evident truths, gradually narrow the fields where possible success is sought, to what we feel as judgment and reflection. In corresponding manner, the contents of expectation differentiate themselves from the simplest mutations up to ideas and world views.

On the side of expectation, the evolution of reason all along its course preserves the creative freedom of true physical chance. Through experience, every gain in knowledge can be preserved for the joint system of expectation and experience, genetically or by tradition. Obviously, the resulting learning consists of an extraction of lawlikeness from the real world, and this can be right only where experience continually corrects the expectations and is therefore within the selection range.

This mechanism contains the history of the Oeser model as far as the realm of the dynamics of scientific theories in all its points. It has "no absolute beginning", contains "no first facts and final grounds", has the character of "reaction", permits "no absolute verification", but, as Erhard Oeser describes, "only alterations of state which are expressed in a constant reconstruction of the theory (the expectation contents), which necessarily inclines to the highest possible stability (probability of survival)". It is an irreversible process with parts of equal value, and which aims to condense information and adaptation, that is to say, conformity with the world that has selected the mechanism.

In contrast, the history of this insight is quite short. Whewell was one of the first to consider the inductive phase as the pacemaker in the development of knowledge. To quote Oeser, "Volkmann, Mach and Ostwald have finally accomplished the last step towards a clear model idea when they characterised this 'self-regulating' process as 'circulation' or 'cycle' with retro-acting 'consolidation'"(11).

The problem of reality

The prerequisite of such an algorithm of cognitive gain is a well-regulated world, with high redundancy. Therefore, it has an order pattern in which regularity repeats itself in comparable application. In a world without order, even in a cosmos of pure regularity without redundancy(12), there would be nothing we could learn with the algorithm. However, it is precisely this order, which maintains, genetically transmits and through tradition hands on its high redundancy, its norms in interdependent and hierarchical patterns, that all sciences seem to discover(13). The problem of the world's reality now looks quite different.

Once again, we can agree with Kant and Popper, who call it a scandal that philosophy is not able to solve the problem. Besides, as Donald Campbell says, we make no "attempt at all to contradict determined (and consequently irrefutable) solipsism. We want to recognise its logical irrefutability"(14). However, it obviously contradicts life, evolution and our reflection on it. If we ourselves, together with our cognitive apparatus, can understand nothing except the product of an evolution that could learn only from the order in the world, then this cannot be less real than the reality of his thought that the solipsist may value.

"Even that we exist and can ask ourselves how our conclusions about regularity in nature can be justified", says S. C. Pepper, "is an overwhelming justification for our belief in this . . . if our immediate surroundings did not possess this regularity, we would not be here to put such questions"(15). George Gaylord Simpson remarked that "the monkey, which had no realistic awareness of the bough was a dead monkey soon after leaping; and therefore did not belong amongst our ancestors"(16). "He who frames a false theory about the world on the basis of his false cognitive categories", to quote Hans Mohr, "will succumb in the 'struggle for existence' — in any case, at the time when evolution of the genus, Homo, was proceeding"(17). Our cognitive capacity is an adaptation to the regularity of the world, says Bernhard Rensch, "the correlate", according to Sachsse, "of what is constant in the environment.". According to Einstein, "even if the axioms of theories are fixed by men, the success of such an enterprise nevertheless presupposes a high level of order in the real world". Therefore there is "really nothing especially remarkable in our ability to make correct predictions about the regularities in our immediate surroundings". If we could not do so then we would not be here "to notice our errors", as Pepper rightly says(18).

There is no longer any question as to how certain the reality of this world is, but only as to how certain the knowledge might be that we extract from it. For, in fact, we can only surmise. Still, some of our conjectures prove so correct that a man can be placed on the moon and brought back again to his family. The certainty is clearly not absolute for the life-span of our cultures, but usually secure enough.

Even the contrast between monism and dualism(19) loses its point. Since it turns out that mind is developed on the laws of matter, the separation or fusion of the two can at best contain the limits of our idea of them, just as space and time appear to us as of two kinds.

The problem of induction

Logic teaches us that there are no truth-extending inferences. Being aware of however many white swans, never compels the inductive inference that all other swans must be white. "The range and force of Hume's argument has been underestimated time and again"(20). Here we must agree with Wolfgang Stegmüller; indeed, we are of the opinion that from however many inferences, say, from the general to the special, we cannot give compelling grounds for inferring from the special to the general. "The step into the unknown cannot occur other than blindly", says Donald Campbell, "and if it is done step by step, then this merely points to knowledge that was already acquired"(21). Logic can only transmit the truth which it possesses but cannot extend it.

We have found that the inductive method has nothing to do with formal logic; for scientific logic, which began as a "doctrine of correct thinking", has retreated to formal logic as a doctrine of the principles of valid argument(22), to the realm of deduction, where binding, compelling inferences are really possible. Logic, since Frege, has banned heuristics, the imprecise and imaginative sister, and seeks to replace her fancy with the precision of the remaining formalism. That, however, cannot be done. The inductive inference does not extend truth but expectation. It has no place in formal logic.

The drives of this expectation do not go back to the reason of logic (they always begin subjectively and illogically) but to pressure from the living; to intentions, goals and wishes (as recalled from Fig. 58), curiosity, appetite and hence to endogenous conditions. David Hume had surmised that wisdom of nature would secure a necessary mental act by an instinct or a mechanical tendency(23). The notions that arise have no compelling necessity, they are "rather", says Albert Einstein, "a free creation of the human (or animal(!)) spirit"(24).

Every expectation nearly always needs to be corrected by experience. Certainly, some swans prove to be black, but this is the next part in the cycle. The uneasiness of the living, life itself, is structured anticipation. Otherwise it would not experience anything. As Donald Campbell says, life is a hypothetical realist.

Even the conflict between determinism and indeterminism(25) is solved; for since living creatures constantly require the creative freedom of inductive conjecture in order to extract the world's regularities, there is no longer any doubt of their indeterminist component.

The problem of the a priori

"The question whether there are synthetic judgments a priori", Wolfgang Stegmüller sums up, "is a fateful question of philosophy"(26). Since Aristotle reflected on "what was proposed on the Agora, the place of judgment", it has remained unexplained whence come these categories of our thought, namely of where and when, quantity and quality, of relation, what and being. Kant analyses them critically as the necessary preconditions of every possible experience. Thus they cannot be founded on man's experience. The foundation of all acquisition of knowledge remains indeterminable(27); indeed, uncertain, for how could a cognitive process be established when from the start it requires knowledge in order to create knowledge?

The doctrine of evolution first brought the solution, most profoundly through Konrad Lorenz, and so provided the foundation for evolutionary epistemology and the possibility of this book. "Our forms of observation and categories, determined before any individual experience", says Lorenz, "are suitable for the external world for the very same reasons as those why the horse's hoof, even before his birth, suits the soil of the steppe, the fins of the fish, even before it is hatched from the egg, suits the water"(28). Therefore the categories are certainly a priori for each individual, but at the same time a posteriori, knowledge acquired through the experience of his species.

This is the solution that Kant himself had conjectured. In his dissertation, he asks whether our notions of space and time are "innate or acquired". After the great critical works he sums up: "however, there must nevertheless be a basis for it in the subject ... and this basis at least is innate"(29). The solution lay in the air; Donald Campbell has tracked down thirty authors who thought the same: Mach and Boltzmann, — Spencer, Mill, Pepper and Popper, — Piaget, Bertalanffy, Simpson and Waddington, — Levi-Strauss, Chomsky and Lenneberg may be cited(30).

The development of the theory of evolution and of evolutionary epistemology, as Gerhard Vollmer classified them, has finally enabled us systematically to solve Kant's a priori. We have derived them from the history of organisms, as a system of hypotheses which evolution has incorporated in the world view of its creations.

This resolves the dispute between rationalism and empiricism as well(31). The rationalists are right, that there can be no experience without reason. We agree with the empiricists in that any reason must rest on experience. Rationalism overestimates the innate experience, empiricism the particular things that can be acquired. In their incompatibility they are both in error.

The controversy between cause and form

It has fallen to modern science to think of the investigation of causes as the real core of its task, but of that of form as not truly belonging to its province. Now we know that experiences of cause and form arise from two sorts of categories; indeed, as we shall learn from R. W. Sperry and John Eccles(32), they are separated in accordance with the hemispheres of the brain. We find that the synthetic processing of form is placed in the right and "silent" hemisphere, whose processes operate unconsciously and offer their results to consciousness. Analytical causal experience comes from the left and "audible" hemisphere and is more readily available to conscious processing. Hence we have long owned mathematics, formal logic and a theorem of causality. However, we have only just had to develop a general theorem of comparisons. Hitherto, the sciences did not have one.

Where attempts have been made with the principles of comparison, as with morphology from Goethe up to Adolf Remane, the process was dismissed from the sciences as "German idealism"(33) and morphology was lost. Its homology theory was cast aside and the type concept rejected. Its field of application, primarily comparative anatomy, then systematics, were grossly neglected, and the whole business of discovery by the growing majority of causal researchers was abandoned to the experimental "exact" sciences(34).

Lorenz had hoped that nobody would "wish to deny the close connections existing between the achievements of creative perception here described and genuine concept formation". For "this process", continued Vollmer, "is nothing else than a preconceptual abstraction"(35). Besides, the causes of type and homology, Goethe's "esoteric principle", could be grasped from the "system conditions of evolution"; and the cause of the perception of form could be clarified and therefore established as a fundamental source of scientific knowledge, as understood by Lorenz(36). Centuries of preconsciously correct knowledge of the morphologists were rehabilitated; and the main theorem of man that is completely rooted in it — the knowledge of his own origin.

In addition to all that, we have demonstrated that cognition of form must be prior to the knowledge of cause; the hypothesis of the comparable prior to the hypothesis of original causes. For, to what should the insight into a cause relate, how could it be imagined reproducible, if it were not first recognised as dealing with the same objects? Clearly, the experimenter can leave the previously required comparison to the processing ability of his ratiomorphic apparatus. Nearly all of them do just that. However, he should not imagine that he could make inferences without knowledge of form, or that his conclusions could be more certain than their premisses. Thus, we are close to the solution of the controversy between materialism and idealism as well. This becomes clearer still in the next problem; we shall discuss it there.

The controversy between causality and finality

Anaxagoras rightly found the materialist interpretation of the world unsatisfying and opposed it with a sense- and purpose-directed world process(37). Next, Aristotle formulated his four causes. His commentators took from these the final cause and, wrongly, made it into the original cause. Since Galileo and Newton, science quite rightly could not do anything with it and wrongly found the original cause in Aristotle's efficient cause, in forces. Since then, the explanation of the world has been split. Those who regard themselves as scientific explain it by materialist causes, those who feel superior to them explain it teleologically, finally and idealistically. There, on the whole, the matter rests. Even with Nicolai Hartmann, who was particularly near to the solution, the seeming opposition between causal nexus and final nexus ultimately prevented the solution(38).

The misunderstanding is rooted, as we saw, in the fact that our healthy common sense inherited too simplified a set of notions as to causes. This latest a priori, that of purpose, Kant dealt with mainly in his last critique, that of judgment(39). We are guided by the view that causes run in executive chains and, worse still, that purposes act from the future; first, because we learn the connection between cause and effect in the performance of our own simplest actions, and secondly because we confuse the accomplishment of a purpose in the future with its premisses, which lie wholly in the present.

On the contrary, we found that causes are functionally interconnected and the distinction between impulsive and purposive causes is that the former act from the less to the more complex layers, while the latter, conversely, act from the more to the less complex; as we experience it, by layers, as selection, breeding, choice, judgment and reason(40). They are all interconnected. We have oversimplified our explanation of the world. "We are playing too easy a game"; as, indeed, Nicolai Hartmann knew.

The controversy between materialism and idealism is likewise solved, however savagely it has ploughed through our history in the guise of ideologies. Each of the philosophies contains half the truth but claims to judge over the whole of the other. Actually, causal relations act upwards materialistically, but downwards idealistically, through the stratified complexity of the real world; dialectically, if you like(41). Therefore, dialectical materialism too is a contradiction in itself; for either it recognises the downward course of causes and then its philosophy is dialectic, or it does not recognise them and then it is materialistic.

WHAT FOLLOWS FROM THE SOLUTIONS

So far we have had to summarise what our theory proposes in the way of solutions. An obligatory exercise, as it were, of one who expects that the value of his theory would be seen in its explanatory force. That is why we have presented solutions where philosophy and epistemology have clearly defined the open questions concerning the cognitive process: in the problems of the trilemma of reality, induction and the categories, and in the controversies about form and finality. With fitting modesty, we have noted that some problems whose solution we present, are as old as philosophy itself.

Here we might end, were there not a number of further problems regarding reason. They differ from the ones mentioned by the fact that they remain rather unformulated. This is not because they are less important; they are even the more relevant to life, as we say today. They are unformulated because it takes our solutions to make them accessible.

Up to this limit of insight, evolutionary epistemology has won significant advocates; since the forties, when Lorenz discovered the background to the a priori; in the fifties, von Bertalanffy and Campbell; in the sixties, Chomsky and Furth, Mohr, Piaget and Rensch; and in the seventies, Lenneberg and Monod, Popper and Oeser, Shimony and Vollmer(42). Here we go one step further, to what follows from the solutions. The trusting reader should therefore be warned. We examine what is the unreasonable in reason; again, from the outside, from the point of view of biology.

This brings us to criticisms on the interaction of conscious and unconscious processes. Indeed from quite a different side to that of Freud, who is a stranger to the scientist, of Jung who is distant from him and Erich Neumann, who comes somewhat nearer to us with his view that ego-consciousness, in the ontogeny of the child, "had to go through stages similar to those which have determined the development of consciousness within mankind". However that may be, we remain on the objective pathway of biology, like Konrad Lorenz(43).

The home of certainty

"Scientific knowledge", said Roman Sexl, "is the victory over the ratiomorphic apparatus"(44). Indeed so, with all the pros and cons. The carefully reflecting consciousness begins to judge its teacher; and, as we shall see, it becomes equally unreasonable when it trusts this background implicitly as when it fully denies it. Thus, we recall, science has for centuries sought that Archimedean point of absolute certainty, triumphing if it thought to have grasped one, despairing when the point remained elusive; whereas mankind, the non-scientific as well as the scientific, beyond all triumph or despair, begat their children, brought them up and handed on their concerns before their life's course was run.

Today even science accepts that there is nothing to be known with absolute certainty. No longer need this "sear our heart"(45). Certainty is one of life's demands and therefore has meaning only within life's span. Obviously this book could cool down suddenly to absolute zero and then fly to the ceiling at relativistic speed(46). Yet mankind can depend on it that as long as our planet lasts this will never happen. Of course, we are surrounded by boundaries to cognitive ability; by the limits of the senses, according to Hume, those of the understanding, according to Kant, of the performance of the brain, added Hubert Rohracher, and of the spirit, concluded Chomsky(47). Optimisation of our cognitive capacity always finds an upper limit, it is imprisoned in the circle of our learning conditions (cf. Fig. 58); even in the contents of our expectations. What sort of presumption would it be if the tick wanted to get some idea of mammalian blood vessels, or a police dog of the international narcotic scene, or we ourselves of the laws beyond this cosmos? Knowledge is only to be understood from the measure of its creature and can conform only with it. "How short soever their knowledge may come of a universal or perfect understanding of whatsoever is, it yet secures their great concernments, that they have light enough to lead them to . . . the sight of their own duties." Thus John Locke(48).

What we experience as the degree of certainty, of an observation or an explanation, nestles at the centre of that far-flung hierarchy of coincidences, which we call experiences, empirical propositions, or natural laws and their cases. In order to feel certain, we need as many downward cases as possible and further sub-cases of these cases; in order to perceive the matter as explained as certain, we need superimposed propositions at the top. The most comprehensive propositions always dispense with explanations since they merely describe coincidences. However demonstrably all knowledge of the world is a system of theories that is stabilisable only from within, as feeling men we shall go on seeking the groves of absolute certainty. Otherwise we should no longer be engaged in the cognitive process of the living.

On atavisms and the emancipation of reason

Particularly at the level of civilisation, man's reason requires both his innate teacher and his conscious reflecting superstructure. It needs the co-operation of ratiomorphic and rational accomplishments. Each by itself makes catastrophic mistakes. This assertion may seem rash, but it is easily verified.

As we have seen, no knowledge can go further than its experiential content. At the same time, it must rest on the precognition from which it has arisen. The world picture of every creature can become appropriate only in the selection area within which its hypotheses can continually fail or be verified. However, no world picture can arise without forerunners, and every extension to the area of life compels the picture and its machinery to extrapolate into the uncertain.

In the first evolution, when only the genetic material was learning, the process was so slow that the world pictures of animals all turned out correct, however small the area of this world may be that is relevant for them. In the second evolution, which handed on experience by means of speech and writing, the process became accelerated by several orders of magnitude. The relevant areas are expanded. The innate teachers are soon overworked. Selection carries out a correspondingly more rapid form of error selection: conscious reflection. This runs the risk of becoming detached from its base.

At that stage, the old teachers of our reason appear quite overloaded. Their only advice in our technological success-bent civilisation becomes an anachronism. It was certainly not made for such problems. This innate adviser was selected for primitive vertebrates, then for mammals and troops of large apes. Now, as we recall, it seeks laws where there are none, form where there cannot be any, it finds necessity more quickly than is possible, it cannot recognise chance as such, it dams up aggressions, remains adapted to optical inhibition mechanisms, presses hither and thither between urges for protection and freedom, it sees causes in the form of chains, expects to be able to find their beginnings, scarcely reckons with feedback but certainly holds that causes could act from the future. It becomes an atavism of reason. We agree with Bertrand Russell that its learning stopped 500,000 years ago. Since then, the innate understanding has increased all(49). So much for the deficiencies of atavism; to those of emancipated reason we shall return later.

The cognitive process and the hemispheres

The development of man was accomplished by conscious reflection, which gallops away from its teachers. This reflection does indeed perceive much, but least of all its own background; and where that emerges, one is ashamed of it, so that it is repressed. However, it can be banished only from consciousness, for it is inheritable. Let us attempt an explanation of this split.

Out of the spiral of the cognitive process only the logical deductive half is rationalised. For oddly enough, it lies in the left hemisphere of the brain, along with the three parts of the speech cortex. Even more surprisingly, only the left hemisphere has a complete link with consciousness. This logical deductive half of the cognitive process is now developed further with consciousness. The left half of the brain thus develops its verbal and analytical abilities in a sequential, arithmetical, computer-like manner, as neurophysiology has discovered. Among the authors concerned we mention Gazzaniga, Sperry and Walsh(50). From this deductive half, men form arithmetic, formal logic and computer technique. Epistemology(51) differentiates it into proof-theory and confirmation-theory with the sequence of deduction, prognosis and reduction. This differentiation of the experiential half in the cognitive process, when compared with its teachers, is almost a new formation through conscious reason (Fig. 59) and this again has dictated the development of civilisation.

In contrast, the old endogenous major parts of the cycle, according to Fig. 58, belong almost entirely to the inductive, heuristic half. They remain in the darkness of the unconscious. These synthetically formed faculties in the expectation half of our cognitive process agree so much with the non-verbal, synthetic-holistic spatial understanding, with the performance of understanding pattern, music and shape, in the right hemisphere, that they must lie in it. Yet this half of the brain has possessed no complete link with consciousness (Fig. 60). Nothing can be experienced of its differentiation; nothing was reproduced. On the contrary, civilisation forced holism out of biology, wholeness out of psychology and heuristics out of logic. Neurologists spoke of an empty hemisphere and still speak of the subordinate one.

We are only just beginning to surmise why only one of our hemispheres is linked to consciousness. R. Sperry had conjectured a complementarity, a working division of the capacities, and John Eccles and others have followed him in this(52). How are we to interpret the unsymmetrical link to consciousness? We suppose, in addition, a connection between early consciousness and its rapidly developing tasks of testing deductively. As may be recalled, these new tasks of man must have been associated with observations of his own actions; with executive learning, first of the actions themselves, and then of ideas about these, and finally of the communication of such ideas, that is of speech. In the rapidly growing space of the relevant, deductive testing foresight could probably claim selective precedence.

However this may become clear in the end, conscious reason now seeks to move free from its innate teachers and thereby gets into the strangest contradictions(53). First, it uses the anachronistic instructions and extrapolates them into a contradiction of a causal-executive versus a final-executive explanation of the world. Secondly, it seeks to emancipate itself completely and reaches a position where it cannot grasp its own knowledge, or establish the reality of the world, or more simply: it reaches the point of mistaking genuine chance for necessity, and conversely. This conscious deductive reason would even make its bearer unfit for life, except that in emergencies common sense takes over and temporarily baffles reason. These are the deficiencies of emancipated reason. To believe pure nonsense is indeed the privilege of man.

On art in science

"Experience clearly shows that there are two categories of men which differ markedly from each other: artist and thinker". Thus the great physiologist Pavlov, and he continued: "Artists grasp reality in its wholeness, as living, indivisible existence. Thinkers dissect reality and break it up into its details. Later, they put it together, piece by piece, and seek to breathe life into it". In a study of form perception, Konrad Lorenz later established that a synthetic talent, perceiving in pictures, may be distinguished from one that is mainly logical and analytical. This is repeatedly confirmed today(54). Very probably this is due to a differential preference of the hemispheres. We can confirm Lorenz, in that these differences in talent are often so pronounced that the complementary candidates no longer understand each other; this can lead even to a deeply rooted mutual mistrust.

Our civilisation, however, has made a clean sweep of these matters. It has decided to grant quite a different value to the logical, rational capacities than to those of the mere poet, dreamer or fantast. It is able to do this because, as we know from J. Bogen, Wladim Deglin and others, the dominance of a hemisphere is not only innate but can be reinforced by education. Our world — as everyone knows — is therefore one of reason; in it things must happen according to reason, which, as consciousness teaches, is based on logic, calculation and reckoning. Since we do not know what else to promote, these form the main subjects of all education. The actual drill, even contrary to some deeply human tendencies, rests in the deductive subjects. Where would mathematics not be a main subject? Where does deduction from the laws of Latin grammar not receive greater emphasis than the synthetic wisdoms of classical poetic art? Are there grammar schools today where art, music and even biology were not reduced to subsidiary subjects? Where might creativity, sensitivity and experience be subjects of instruction? Does not the reason of our day show that these permit neither formulable tasks nor duties(55)?

What happens in universities? Analyses are taught in all the sciences. For how could synthetic thinking be taught and tested? On the contrary, the consensus of reason has established a further scale of value which leads from bare descriptions, through experimental work to the exact sciences; to the ultima ratio of axiomatic deductive systems, which, if they are absolutely certain, have equally certainly nothing to do with this world. Research promotion is compelled to play this game(56). The variegated background of the world is lost, development becomes education and education, training(57). Even the measured appraisal of human intelligence is reduced according to need, to verbal smartness and calculating cleverness. For how is one to measure the product of creativity and motivation?

A direct consequence of these respectable arrangements is the lack of any agreement between success at school and success in life. There soon follows the specialist's "paean" with a deep disinclination to re-thinking or innovation, and a type of trained person popularly and irreverently called "professional idiots".

Where such a group dominates, a deepening uncertainty causes individual responsibility to lapse, and truth will be decided by majority voting. One ends with decisions which, as we know from Dörner's experiments(58) or from the evening news, must lead to the ruin of all complex systems. Art has forsaken science. It even runs the danger of forsaking itself.

Collective liability and collective error

This is the point at which, praise God, we release the individual from the scruples enumerated. Sociology of knowledge tells us that even reason must be understood as a product of the collective. We individuals play too easy a game, not only because our dispositions teach us this but even more because all the world around us plays the same game and those who do not conform are as ever punished exemplarily. The necessary stability of a culture depends on consensus and on this being immunised against every possible contradiction(59).

Only now does the effect become transparent, which must arise if the collective extrapolates innate teaching into areas of life for which this was simply not created. However, preference for simpler solutions, executive causality with beginning and chain formation, which considers even final causes as opposition, must lead to a collective misappraisal of the human world. As regards learning processes, this erroneous simplification is then expressed uniformly and consistently in the applied forms of ontological reductionism and in the tabula rasa point of view(60). If we keep in mind that the contents of expectation act on the contents of experience, these on the mode of experience and the reaction to this, and further on the impulse of expectation, on the aids to decision and back again to the contents of experience themselves, then we see how difficult it must be to break away from fixed expectations. To the collective mind this seems quite impossible. Consequently those applied forms of ontological reductionism support each other and become invulnerable: positivism, social Darwinism, behaviourism, phenetics and dogmatic genetics all alike(61).

Such collective error is by no means a matter of other-worldly disputation. It is rather a consequence of the success of our successful civilisations and reacts on it as their stylised theory. It now follows that one will fail to see the reactions of industry on markets, politics on the voter, market on capital and equipment, and back again to industry; although Forrester, Galbraith and Jouvenel, Schumacher and many others have made all this quite clear(62). We learn that industry can live only from growth, "power" (i.e. energy, might or capital) only from aggrandisement, ideologies only from their spread. We know this cannot come to a good end; so we look for alternatives, and are surprised that, even if we do find them, they usually turn out to be inapplicable.

So we imprint irreversibly, with status and consumption, flags and uniforms, incantation and oracles, with manipulation and pictures of the enemy, right on to "brain-washing"; the highest form of "perversion" among human learning processes, as Franz Seitelberger has convincingly formulated(63). The environment is reduced and prepared correspondingly, so as to permit all this; to a kind of cultural parasitism, in which alleged certainty runs wild at the expense of individual freedom, and quantity of life at that of quality(64).

Now we biologists need have no worry that the biosphere might no longer regulate itself. It has always managed to cope with all the species that have proliferated; it has selected them. In societies as large as our civilisation, however, it does not select unsuitable individuals, but unsuitable civilisations. The growth of empires in the world's history up to such sizes that they collapse again, seems to be the consequence of rational extrapolation; but the survival of individuals in the ensuing chaos seems to result from their ratiomorphic performances. This constitutes collective liability, which every one of us has long since adopted as the collective mischief of our reason.

On scientific humanism

This was the point in the affliction of civilisation where many now conclude their subject. We evidently do not. As biologists, we do know that our large brain has developed exactly like those extreme organs which, so far, have brought all their owners to extinction. However, we also know that our extreme organ enjoys a certain advantage, which we may have to exploit: the organ can be aware of itself.

So we return to the deeper purpose of this book. What is involved is man and his closer determination; not a mere academic argument, but a more profound understanding of our origin and possible future; hence an improved prospect of what is possible or impossible for us. Granted, this purpose may have arisen from mere curiosity; but this can lead to a part of that gain in knowledge, the course of which we have just described. To the biologist, this step in biology seems like the continuation of the process of cognitive gain, wherein every living creature is trapped in constant movement throughout its life: in a restless search for more vision and foresight; with the unattainable goal of rest and certainty. The purpose is thus both natural and human; to be able to understand and explain why this reason so often plays bad tricks on us. It is therefore the second compulsory exercise of this summarising chapter. It should help us to make use of what is reasonable in our reason. How much would be achieved by this! Evolutionary epistemology can become a superstructure over the superstructures of creative learning; a continuation of the evolution of the evolutionary process(65).

This demand to learn whence we came is certainly legitimate and a demand of humanity. Yet is leads us over hurdles, as Vollmer called the "Copernican revolutions", which in their course resulted in the opposite of humanity; the revolution of Copernicus for Galileo, the second with Darwin for Ernst Haeckel(66). Given reason, may a third one be spared us. For humanity equally rests in quite a different insight. We take this from Albert Schweitzer and change it round for our purpose . . . "we are life that seeks certainty, along with other lives that also seek certainty"(67).

Such requirements of humanity were claimed by movements which we know as political and philosophical humanism. Hegelians have formulated "a historical process of reason for freedom"; Julian Huxley "an evolutionary humanism"; Pope Paul VI demanded in "Populorum progressio" a humanism of peace for peoples and society(68). We feel sympathetic to this notion of a second enlightenment, a liberation from manipulation and corruption by insight and knowledge.

Of course, Marx and the "Monist league" and Sartre have done the same. Humanist socialism became revisionism and so, according to Mao Tse-Tung, the symbol for "permanent revolution". On that account we value our scientific point of view, our objectivity postulate and the claim that experience may frustrate us. Humanity, as we understand it, should help man to make himself free, even against materialism and idealism. It should bring his natural demands to his own attention, in spite of, indeed against, all the half-truths of ideologies, against manipulation by technocracy and against all the incompatibilities of their claims to correctness. We are convinced that in the end only man's objective knowledge will be recognised as that reconciling tribunal which we can call upon, as indeed we must, when ideologies conflict.

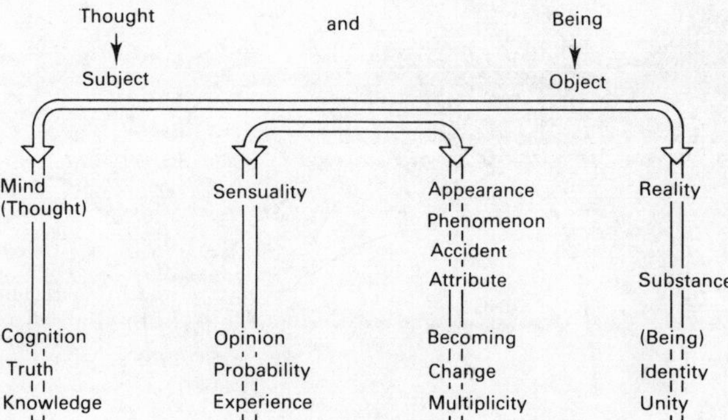

FIG. 1 The "Parmenides Model" is considered as the first philosophical interpretation of the relation between subjective and objective world. This system of the Eleatic school from the fifth century B.C. "has remained essentially valid up to the present day, inasmuch as current explanations still use it as a guide" (from Diemer and Frenzel, 1967, p.36).

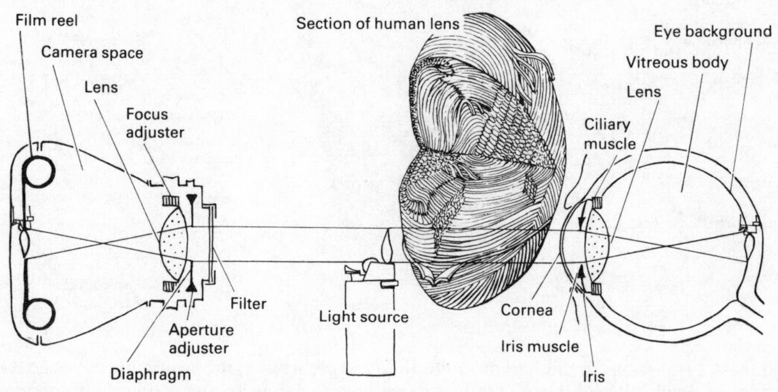

FIG. 2. The eye and the camera reveal largely similar functional components in the two optical systems. A sectional illustration of the lens has been added to show its structure of transparent fibres.

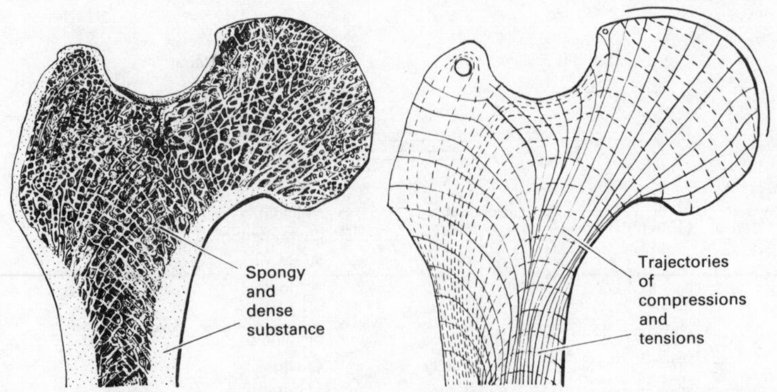

FIG. 3. Bone process and force trajectories as exemplified in a longitudinal section through the head of a human femur. Note the substantial agreement between the position of the process and the compression and tensile stresses in a similarly loaded plastic model (from Toldt and Hochstetter, 1940; Kummer, 1959).

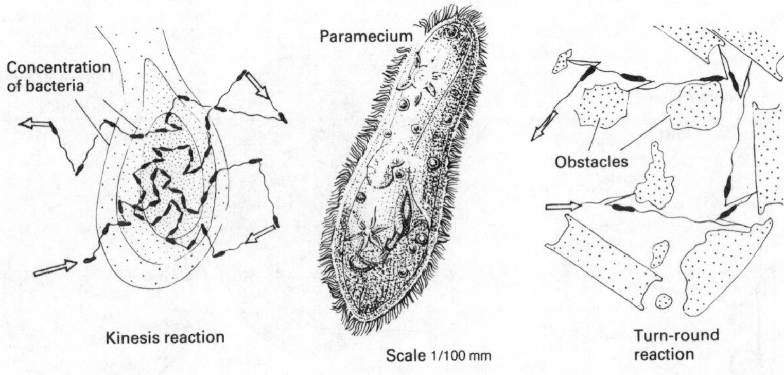

FIG. 4 Reactions in the Paramecium. In kinesis reactions, the swimming movement is automatically slowed down with increasing bacterial density and residence in the food region is thereby extended. The turn-round reaction, in spite of its stereotyped nature, again leads to the successful avoidance of all obstacles.

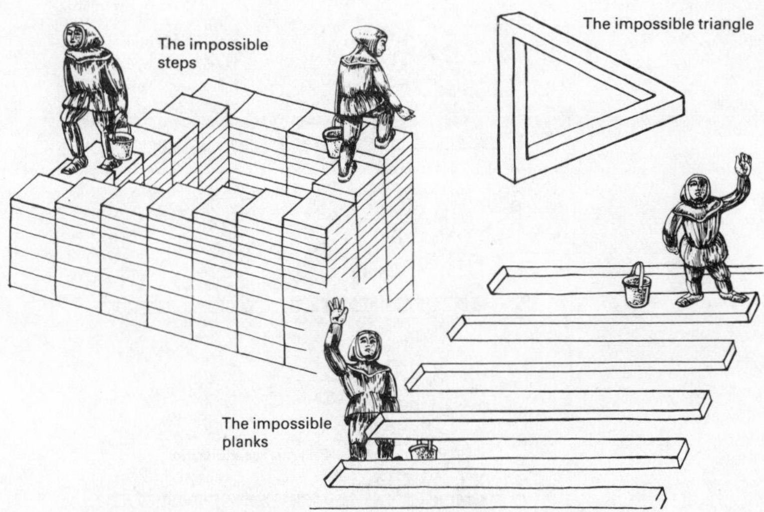

FIG. 5 The "Impossible Figures": because of our inherited space interpretation, we are not able to think of them as possible objects, however well it is clearly possible to draw them (excellent examples are given in Escher, 1975).

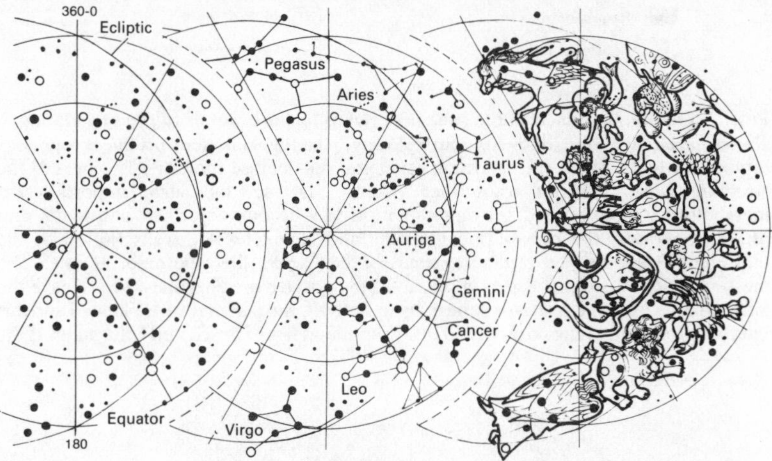

FIG. 6 Form, where none exists, exemplified by a section of the northern celestial sky. One should note the really chance distribution of the stars, their magnitudes and, in the centre, their association into constellations; to the right, the baroque representations according to the sky atlas of Andreas Cellarius, published in Amsterdam in 1708. It is rotated through 90° and turned around the axis 90-270°, so as to correspond to present day orientation (cf. Störig, 1972).

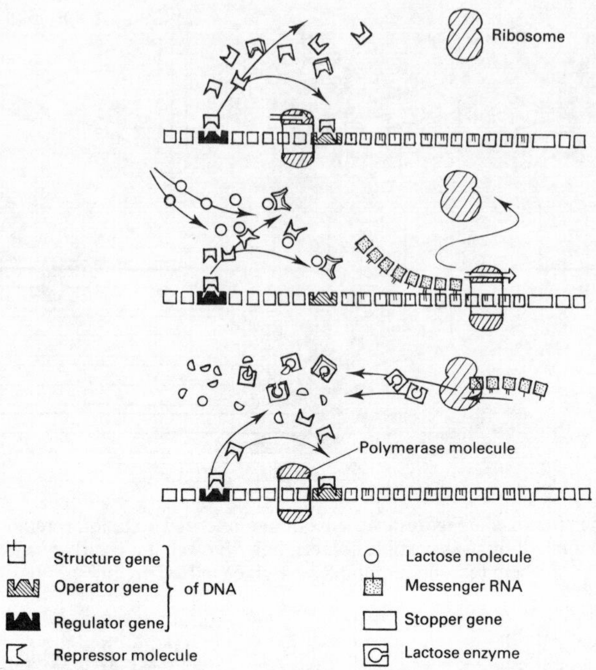

FIG. 7 Learning products in the gene material using the example of gene regulation in *Bacterium coli* with the following course. Above: the regulator gene produces repressor molecules, these block the operator gene, the polymerase molecule is blocked, the structure genes are not transcribed. Centre: lactose, the sugar important for metabolism, has penetrated, the repressor molecules become deformed, and do not shut off the operator, the polymerase migrates and completes the transcription of the messenger-RNA by the structure genes, which seeks the ribosome. Below: The messenger-RNA is transformed in the ribosome into lactose enzymes, these cleave the lactose molecules, the operator gene is again shut off, the polymerase is blocked and this enzyme production is again switched off. (Simplified from Bresch and Hausmann 1972; Watson, 1977).

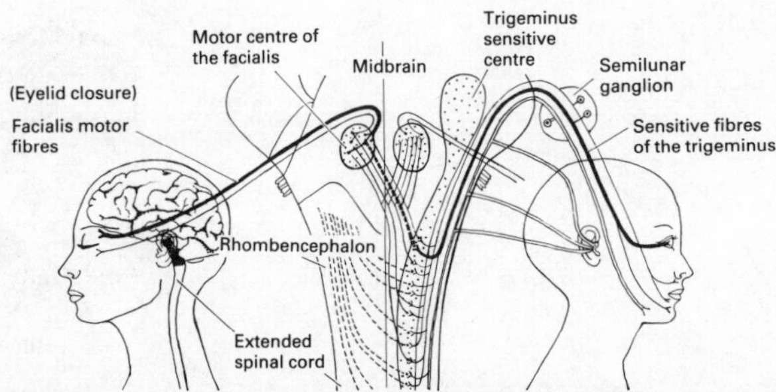

FIG. 8 The conduction of an unconditioned reflex exemplified by the eyelid closure reflex in humans. The probable fibre course is made clear by the pathways drawn in thick lines. The area between mid-brain and the extended spinal cord wherein the operating connection lies, is dark in the left-hand figure and drawn enlarged in the centre (after Crosby, Humphrey and Lauer, 1962).

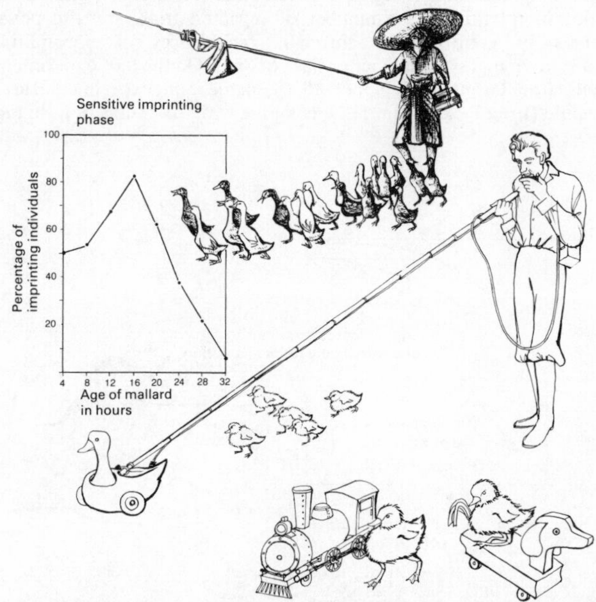

FIG. 9 Imprinting. In Bali it has been the custom for a long time to lead geese to their feeding place and to the stall under their "little flag". This connection was recognised by Konrad Lorenz by using dummies with loudspeakers from which their own voices sounded. Today, the most sensitive times for imprinting have been well investigated and the most absurd imprinting objects are known which the imprinted chicks will follow for the rest of their lives (from Hesse, 1959; Allen, 1972; Heinroth, 1974; and a television film by H. von Ditfurth, 1978).

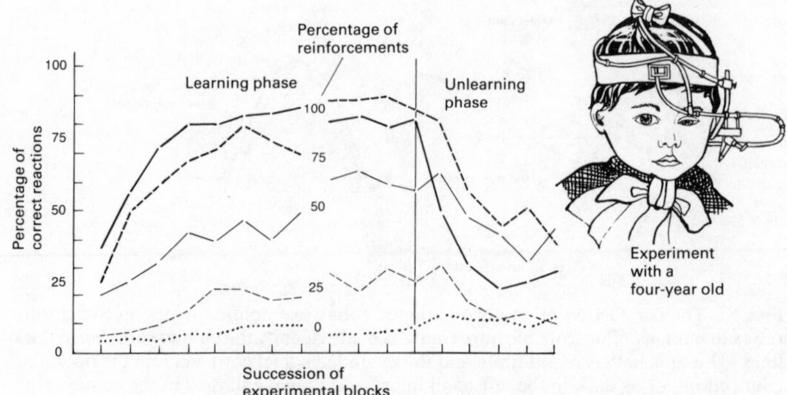

FIG. 10 Learning and unlearning of a conditioned reflex exemplified by the eyelid closure reflex in relation to the number of repeated trials and the percentage of reinforcements or confirmations through coincidences of unconditioned and conditioned stimuli; e.g. stream of air or flash of light. On the left, experimental results from students (from Grant and Schipper, 1952), on the right, experimental arrangement with a child (from Pickenhain, 1959; compare with the connections in Fig. 8).

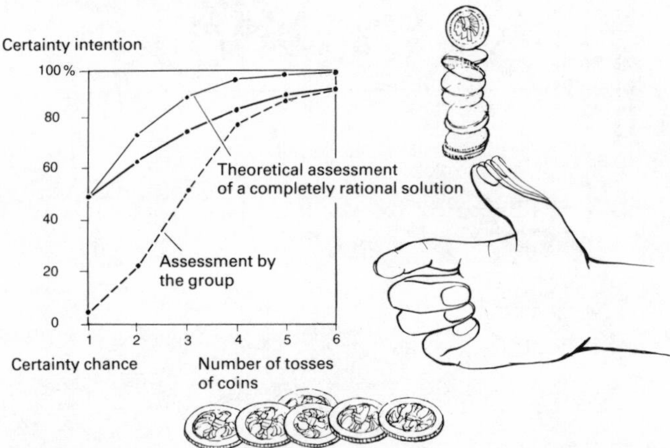

FIG. 11 The conditioned discovery of necessity or of the presence of intention, respectively. 109 students were assembled in the lecture room for an experiment on coin tossing; and whilst only "heads" always turned up, as we who were conducting the experiment had "fixed" it, they had to write down, after each result, how they judged the experiment. As a comparison with this, the critical finding of a "completely rational solution" is given (uncorrected values in thin lines, corrected ones in thick lines).

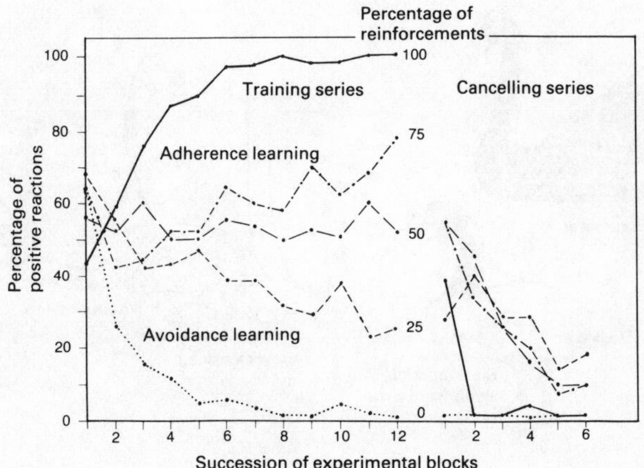

FIG. 12 The effect of reinforcement and disappointment in learning to distinguish between positive and negative reactions; in relation to the duration of the experiment and the relative frequency of reinforcements and disappointments; in this case, of illumination or non-illumination of a lamp after the appearance of a starting light. Note the constant occurrence of uncertainty (50%) with equally frequent disappointments and reinforcements (from Grant, Hake and Hornset, 1951).

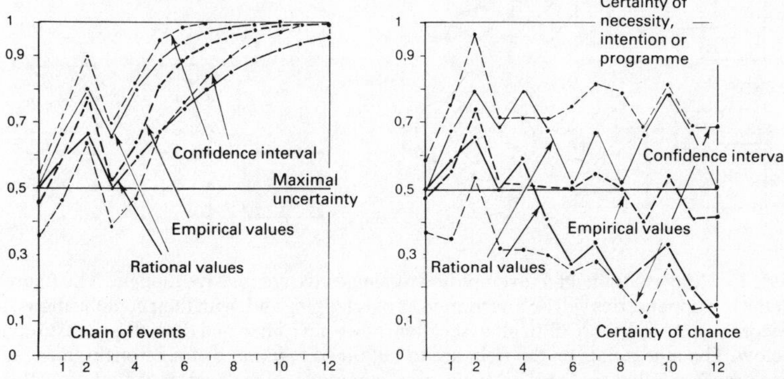

FIG. 13 The growth of the degree of certainty in the stepwise discovery of a programme (left) and a chance series (right); as entered below the diagrams. The chance in judgements of 20 students at a time is shown by the broken lines (mean values, thick; confidence interval, thin lines); the calculated value is abstracted as the judgement chance of a "completely rational person" (corrected course, thick; uncorrected, thin lines; compare also Fig. 11).

FIG. 14 Species of swans; the European species are contrasted with those from the southern hemisphere (left); with the Australian black swan and the black-necked swan from Southern Chile and Tierra del Fuego (from Peterson, Montfort and Hollom, 1954; Klös and Klös, 1968).

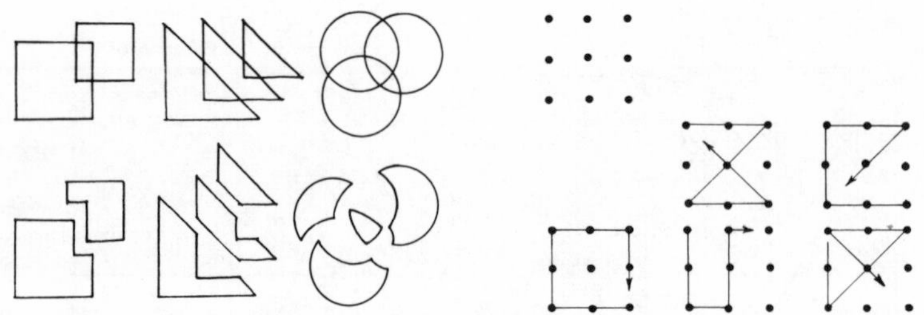

FIG. 15 Magical thought as exemplified by suggestive compulsive thought. The figures in the left upper series will be interpreted as overlapping and, with their complications, it becomes more and more difficult to see them as we have imagined them to be constituted below. The nine points on the right are to be joined together by four straight lines into one drawing. Below that are shown the most common wrong attempts at a solution; they follow the form suggested by a square (correct solution in note 117 to Chapter 2).

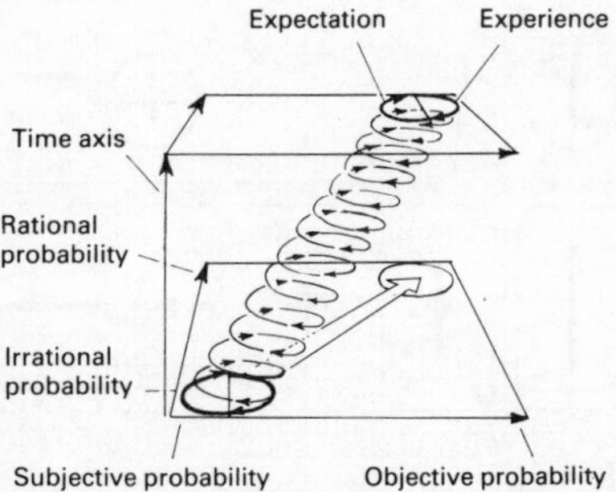

FIG. 16 Alterations in probability in the example of the cycle of expectation and experience. The spiral form alters with time to an extent that, with new experience, the expectation changes and with altered expectation, the experience determining it. The expectation corresponds to a probability which, in the case of complete ignorance, must emerge from a purely subjective and highly irrational (or illogical) form but which, with optimal learning success, can arrive at an objective and rational (logical) form of probability.

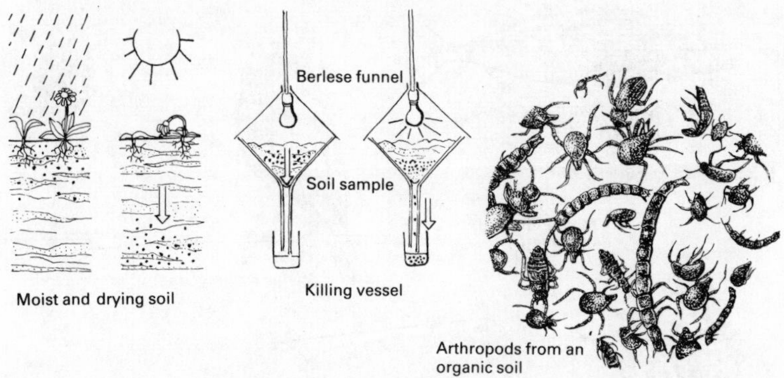

FIG. 17 Reasonableness and absurdity of genetic programmes as shown by examples of soil animals. In nature, when soil dries the animals react positively geotactically and thus reach moister soil layers. If the soil sample is dried under a lamp over a grating then, in conformity with the same programmes, they fall to certain death in the collecting vessel (example of a collection sample, from Kühnelt, 1961).

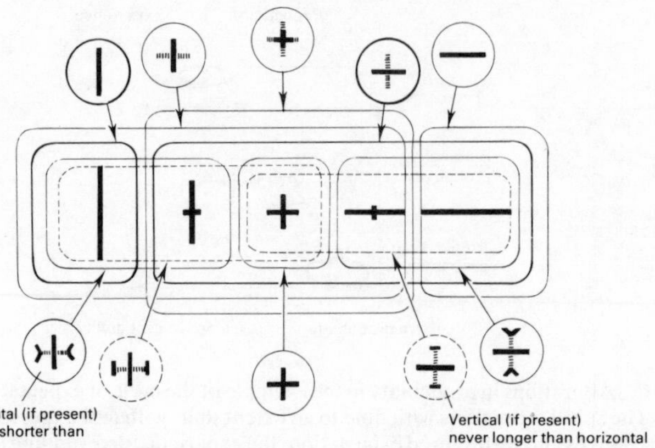

FIG. 18 Flowing similarities and "mapping concepts" intersecting each other. The central five figures, with only two varying characteristics, determine ten units or sub-fields (according to Hassenstein 1954 and 1976). In the circles, the "type features" are represented symbolically. Systematic "definitions", too, are mapping or injunctive concepts of this kind.

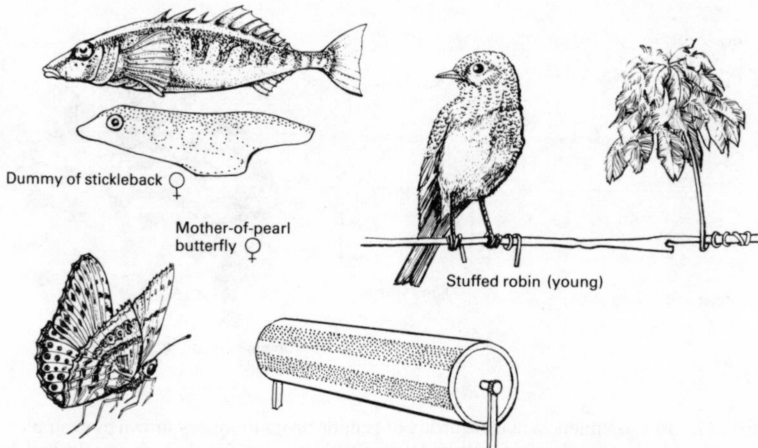

FIG. 19 Super-optimal optical releasers as exemplified in each case by two dummies. Thus, the greatly simplified imitation of the female three-pointed stickleback with exaggerated spawn-carrying abdomen is preferred by the male to the genuine one; the robin prefers the spray of red feathers rather than the stuffed young bird which is not red; and the male emperor mother-of-pearl butterfly prefers a rotating roller with the yellowish brown colour of the two sides of the wing to the normal female (from von Ditfurth, 1976).

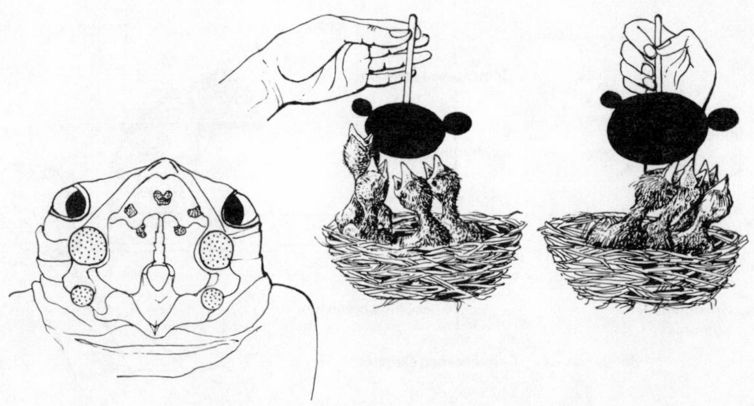

FIG. 20 Innate optical releasers exemplified by signals and signal perception in young birds. On closing the beak, as here in the young Gouldian finch, the black and bright blue signals for the feeding parents become visible. Young blackbirds close their beaks even with the simplest dummies, but they have a very accurate "idea" of proportions so they beg at the head and not at the body of the parents (from Tinbergen and Kuenen, 1939; Eibl-Eibesfeldt, 1978).

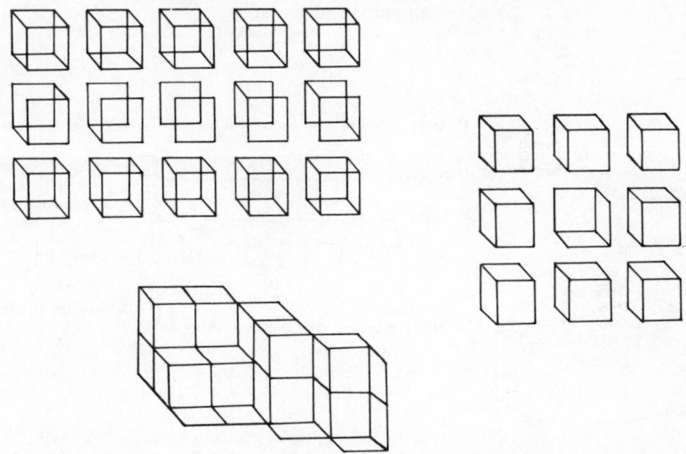

FIG. 21 The spatial interpretations of patterns. The cube-like figures constantly change together between the two possible perspectives. Accordingly, in the central figure of the upper group, one of the squares will always appear to be floating. Below left, the steps permit two equally valid interpretations. The centre figure in the group on the right more often appears as a hollow angle than as a cube because the position of the surrounding figures influences the layout.

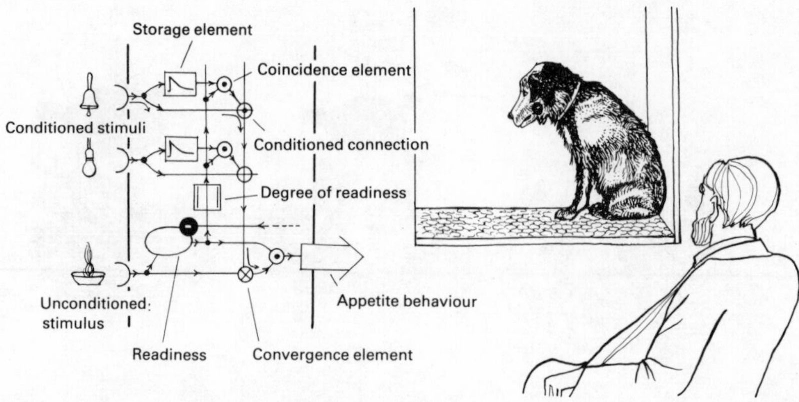

FIG. 22 The connection of a conditioned reaction. In the idealised connection diagram, the arrow shows the direction of the communications running through the nerve paths and the linkage of the conditioned stimulus with the unconditioned stimulus. The change in the degree of readiness reports the appearance of the "reward" to the learning system (from Hassenstein, 1973). Next to it is an extract from the experiments of Pavlov, who described the conditioned saliva reaction of his dog as a conditioned reflex (from Allen, 1972). Today we know that a more complex reaction is involved, namely conditioned appetitive behaviour.

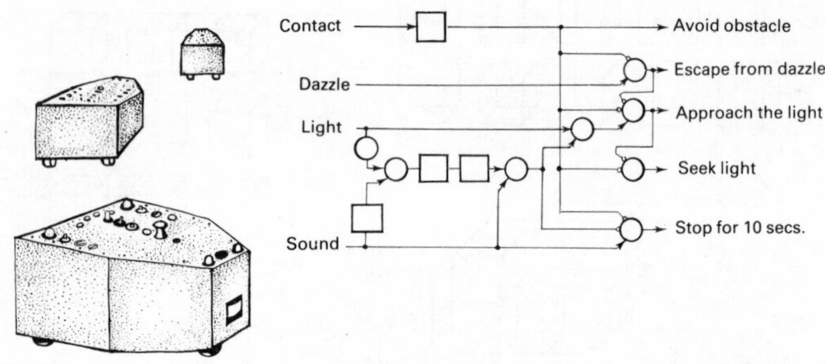

FIG. 23 Conditioned reflex in technology, illustrated by the Vienna "artificial tortoise", an automat for conditioned behaviour. Alongside, diagram of the connections in the first model of conditioned reflex CORA (conditioned reflex analogue) for electronic simulation (cf. Walter, 1951; Zemanek, 1962; Goldscheider and Zemanek, 1971).

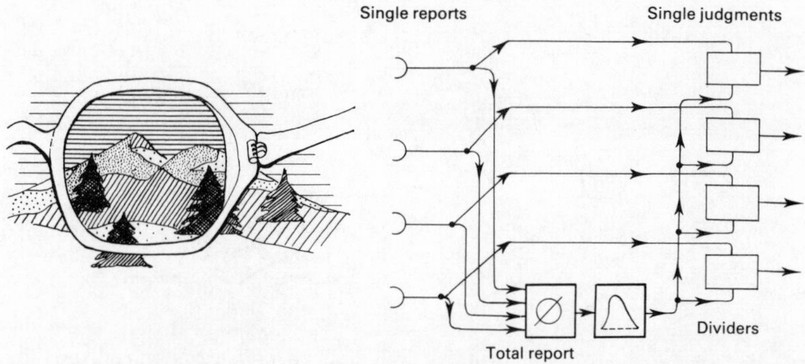

FIG. 24 The invariance performance or constancy perception. In spite of dark sun glasses, for example, after a short adaptation colours as well as highlights will be seen "correctly", although they are wrongly reported. Next to it is the circuit diagram, in which each individual report is assessed against the overall report. This circuit arrangement is used in this form in technology (e.g. a Zeiss polarimeter) (from Hassenstein, 1965, cf. also Sachsse, 1971).

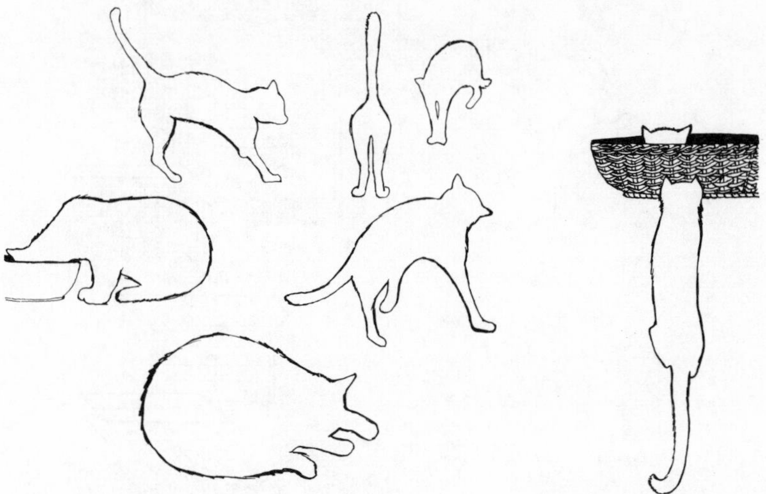

FIG. 25 Constancy perception of form. Although the retinal images of the drawn figures are readily distinguishable, their consideration by abstraction and supplementation in form perception nevertheless leads to the conclusion that a similar, indeed the same, form is involved in all cases. Every representation turns out to be completed from the whole background knowledge about the object.

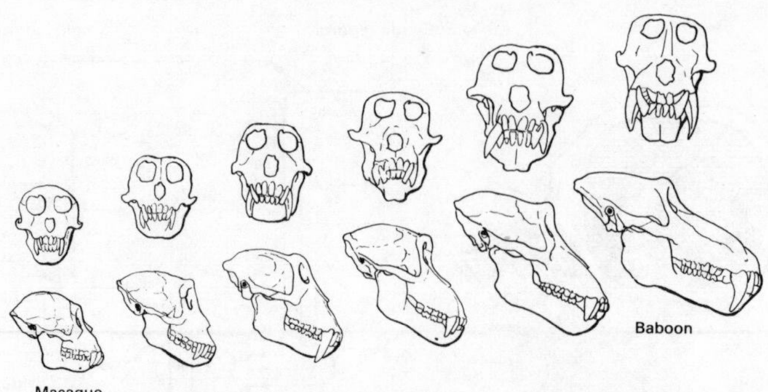

FIG. 26 Homology from a closed divergent similarity field. In an example of this kind, the abundance of feature and form are reinforced in the sense of simultaneous and successive coincidences and so permit no doubt that the representatives of the similarity field can be homologised. Neither external causes nor chance need be considered as explanations of the similarity (after Gregory, 1951).

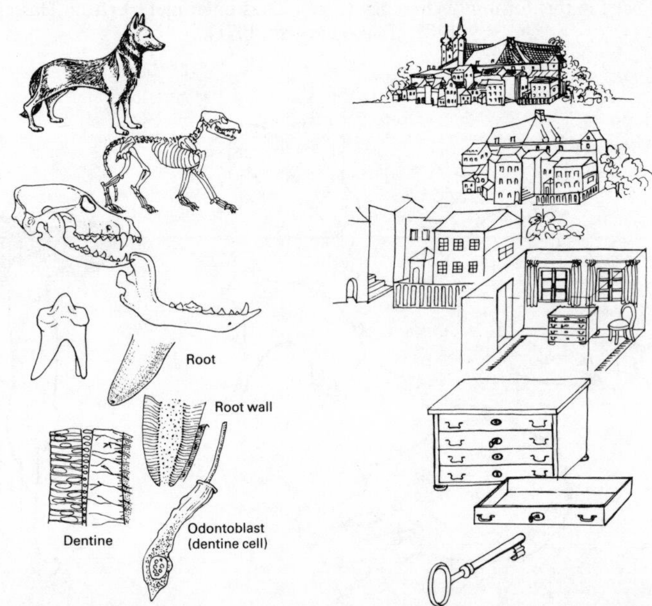

FIG. 27 The hierarchy of all form. As a reminder that a form is always to be considered the form of a further form, it is here shown as a chain of sub-units, since they altogether have or are expected to have meaning only in their higher units. If one assumes that each form consists of ten sub-forms, then with six to nine hierarchical levels, we are dealing with a million to one thousand million sub-forms of a form.

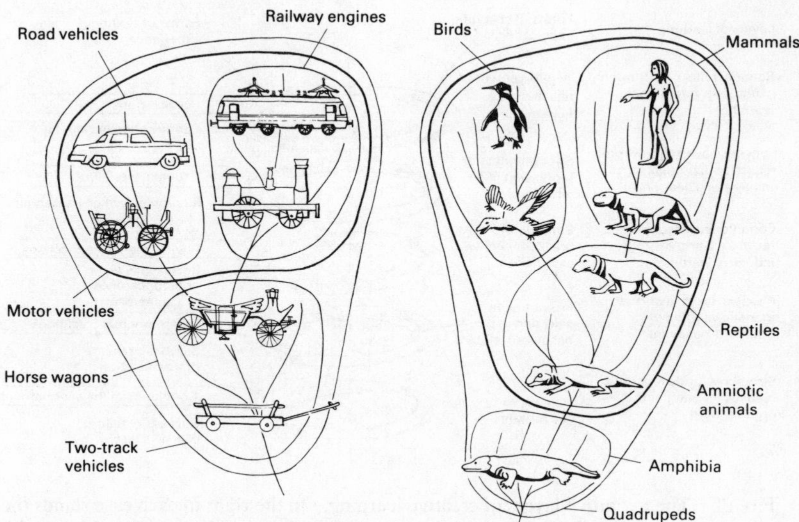

FIG. 28 The hierarchy of all similar fields. The organisation of similarities within a field is alternately determined from the total content of the higher fields just as the classification in the latter emerges from the structure of their sub-fields. Indeed, in our total knowledge each experience is embedded in and composed of such experiences.

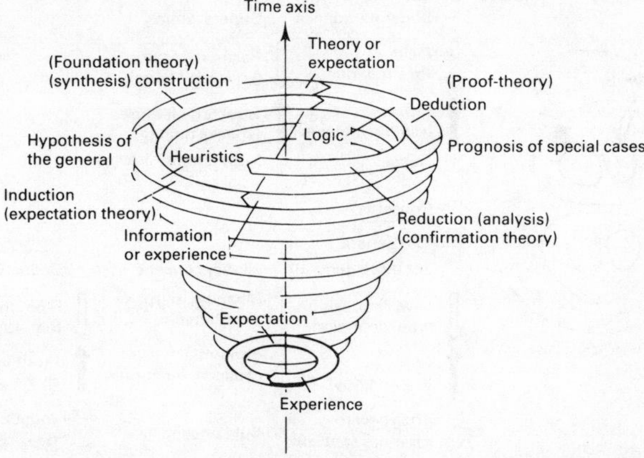

FIG. 29 The cycle of cognitive gain, hence the growth of knowledge and certainty according to Erhard Oeser's theoretical system-functional model of the dynamics of theories. The symmetries contained in this algorithm correspond to those which were found prepared in the phylogeny of biological cognitive processes. It is only that they are more differentiated at the level of epistemology (from Oeser, 1976; extended into biological history).

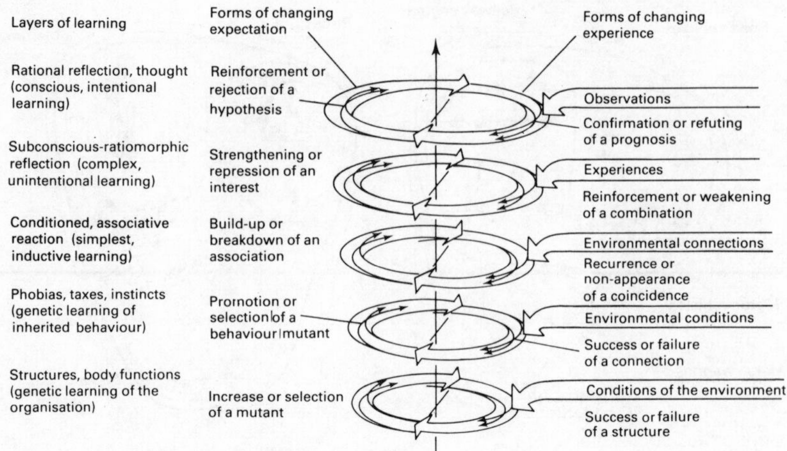

FIG. 30 The evolutive layers of creative learning. On the right in each case stands the experience gained in the most recent past, on the left, the expectation derived from it for the immediate future. The forms of expectation and experience change from layer to layer. The principle of the algorithm remains unchanged, since the formation of each layer presupposes the success of the preceding one (cf. Figs 29 and 58).

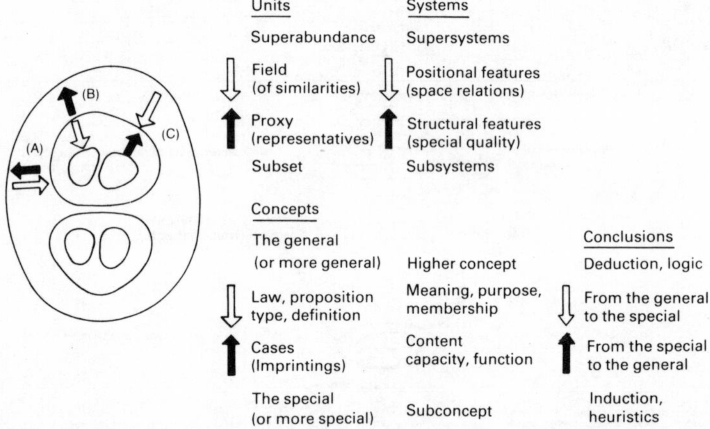

FIG. 31 The regular connection between higher and lower systems without considering the time axis. There are hierarchical connections of higher and lower orders, we can also say of expectation and experience, which are reflected in our notions of the unity and systems of nature as well as in our concepts and inferences. The relation symbolised in (A), (B) and (C) are formulated in the text (compare the consideration in the time axis; Fig. 30 as well as Figs 29 and 16).

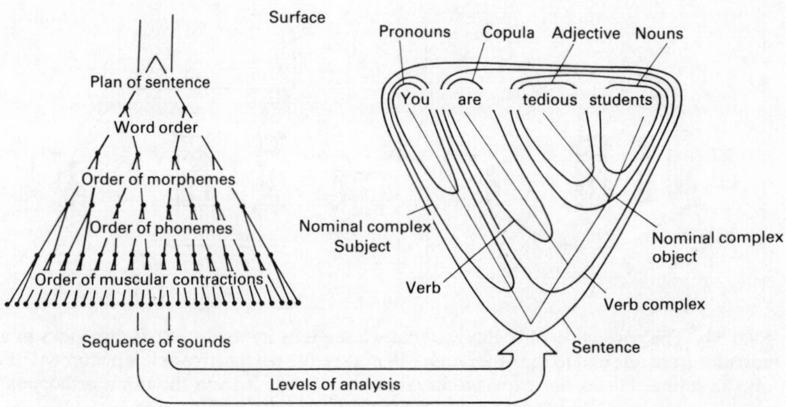

FIG. 32 The hierarchy of language. In the development of speech, all decisions follow from higher decision; in the analysis, the whole is stored, and the sentence interpreted from its parts and these from their morphemes, furthermore the phonemes and sound groups are interpreted in order hierarchically to constitute from them the sense of the sentence at the surface (after Lenneberg, 1967).

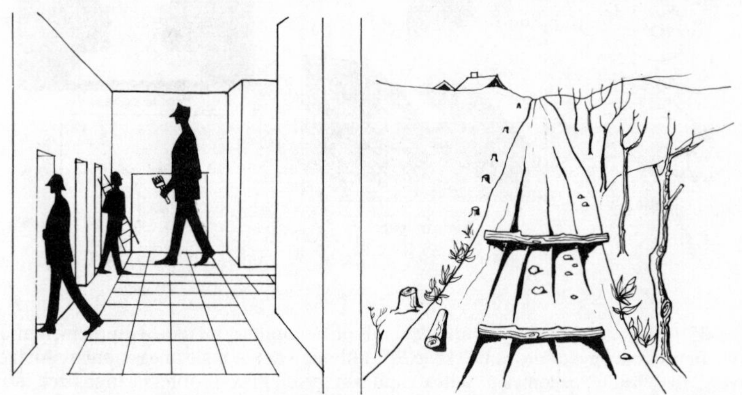

FIG. 33 The so-called perspective illusions, in nature, are corrections of significant life-preserving function. It would be dangerous to underestimate the giant or the obstacle in the background of the pictures just because they are still in the distance. Of course, in the drawings the scale may convince of their equality of size with the figures in the foreground (from Hubert Rohracher, 1971; E. von Holst, 1969).

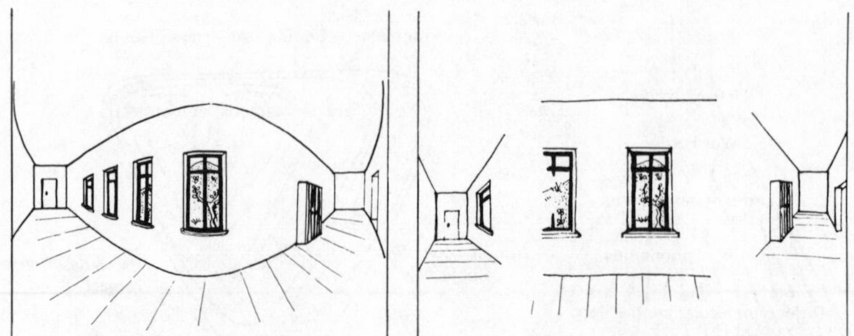

FIG. 34 The correction of reality, so that we see it as it "really" is . If one looks in a corridor from one end to the other, one will not see the perspective as it is pictured (left) on the retina, but as our expectation (right) interprets it from the three orthogonal dimensions of space (from von Holst, 1969).

FIG. 35 The interpretation of form depends on involuntary imagined supplementation. Two figures become of the same "species", although not a single line is similar. Alternatively, the "bride" becomes a "witch" and conversely; two profiles change back into a vase (from Wellek, 1955); and in the extreme case, the doodle leads to extreme completion in jokes ("housewife without a free hand closing the refrigerator").

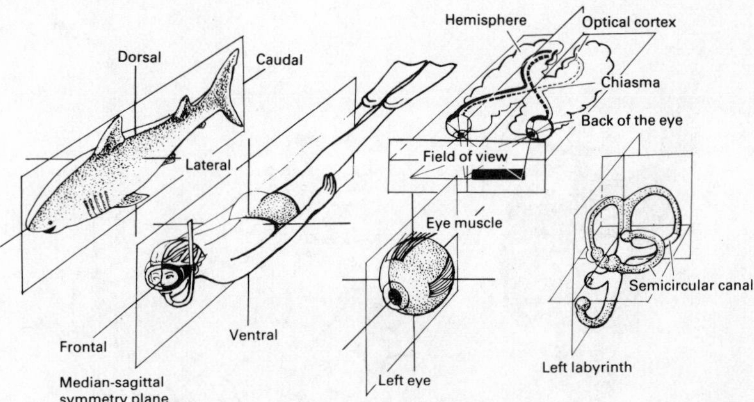

FIG. 36 The biological geometry of our space. The bilateral symmetry of vertebrates corresponds to that of three axes which we accept as those of Euclidean space. For the whole body is organised according to it, including the brain, the higher sense organs, the axis of rotation of the semicircular canals, as well as the connections of the optical pathways. In binocular vision the left-hand fields of view are conducted to the right-hand sight cortex and conversely (adapted from Hochstetter, 1945; Abderhalden, 1946; Romer, 1966; Hubert Rohracher, 1971).

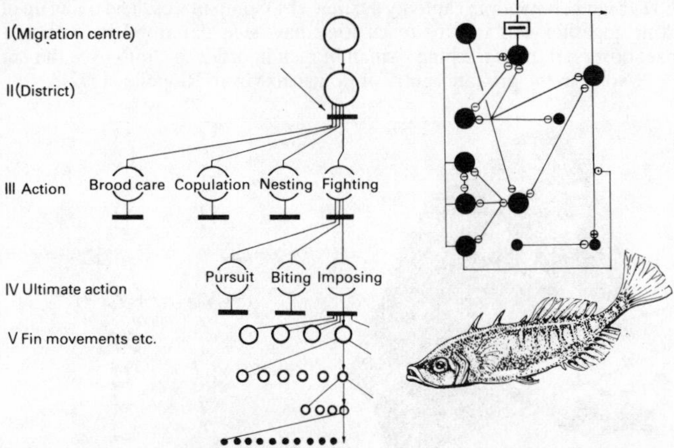

FIG. 37 The hierarchy of instincts shows the further conduction of instinctive actions according to the hierarchically arranged inborn release mechanism; here it is exemplified by the alternative actions and decisions up to the imposition behaviour of the stickleback; the arrow indicates the position of the existing blocks (on the left, greatly simplified from Tinbergen, 1951; on the right, the scheme of interlacing, according to Wickler and Seibt, 1977).

FIG. 38 Design behaviour in captivity as shown by chimpanzees. The piling up of boxes or stacking together of canes to reach food has been described by Köhler (1921). E. Menzel observed them fetching a small branch in order to climb over the enclosing wall, in the primate centre of Louisiana (from Riopelle, 1972).

FIG. 39 The understanding of the "if-then" symbols as illustrated by the chimpanzee "Sarah". The animal is taught to associate plastic shapes with special concepts. Two sentences are cited from the "conversation" which show that the animal understands the token for "if-then" and uses it correctly (according to Premack, 1971, from Riopelle, 1972; in addition, see the clear discussion in Watzlawick, 1976).

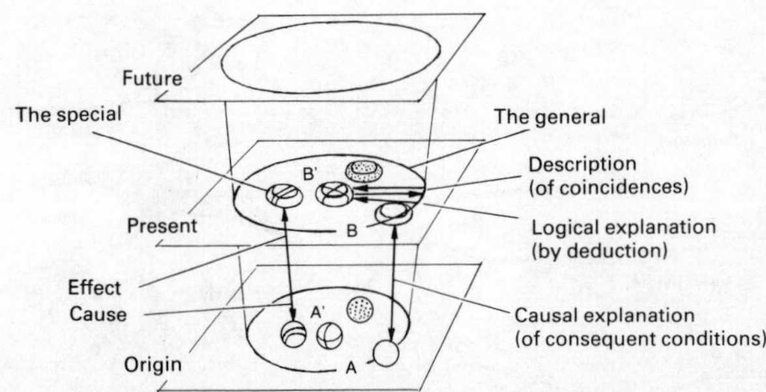

FIG. 40 The connection between explanation and description and between cause and effect. Description is a prerequisite for explanation if something is to be shown about its sphere of validity. On the left, the time relation; on the right, the cognitive relation of the concepts (compare also, Figs 31 and 46).

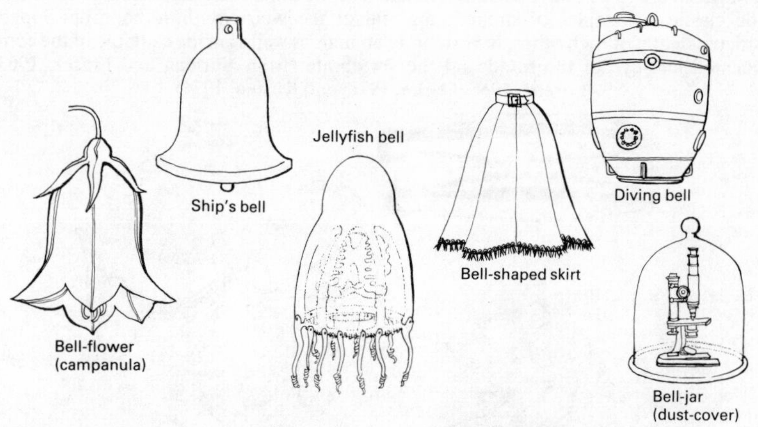

FIG. 41 At the limits of chance analogies. For example, the repetition of the bell shape can be explained in only a few cases as a recurrence of the same cause; much more likely the same effect will be attributable, by chance, to different causes.

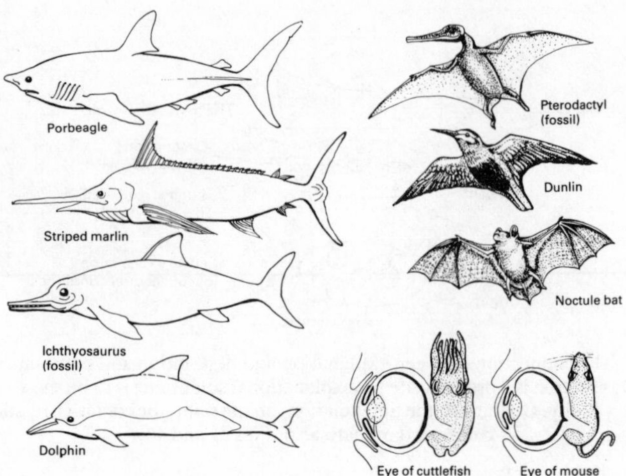

FIG. 42 Functional analogies we explain as similarities emerging from different forms and which are to be understood as a reaction to the same external conditions. Here, in the classical examples of streamlining and of the wing, as these have developed, independently of each other, in vertebrate animals as well as in the example of the complicated lens-eye of the squid and the vertebrate (from Norman and Fraser, 1963; Lorenz, 1965; Osche, 1972; and Kurten, 1974).

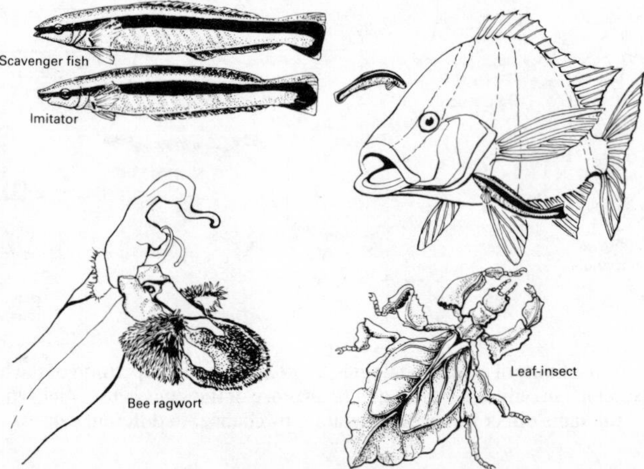

FIG. 43 Mimicry, an extreme form of functional analogy, demonstrates imitation and deception in nature. The predatory imitator of the scavenger fish, for example, by its disguise can insinuate itself with large fishes. The flower tips of ragwort act as a dummy female bee and attract the male for copulation, and thus serve for pollination. The "walking leaf" proves to be a stick-insect optimally camouflaged in the foliage (after Wickler, 1968).

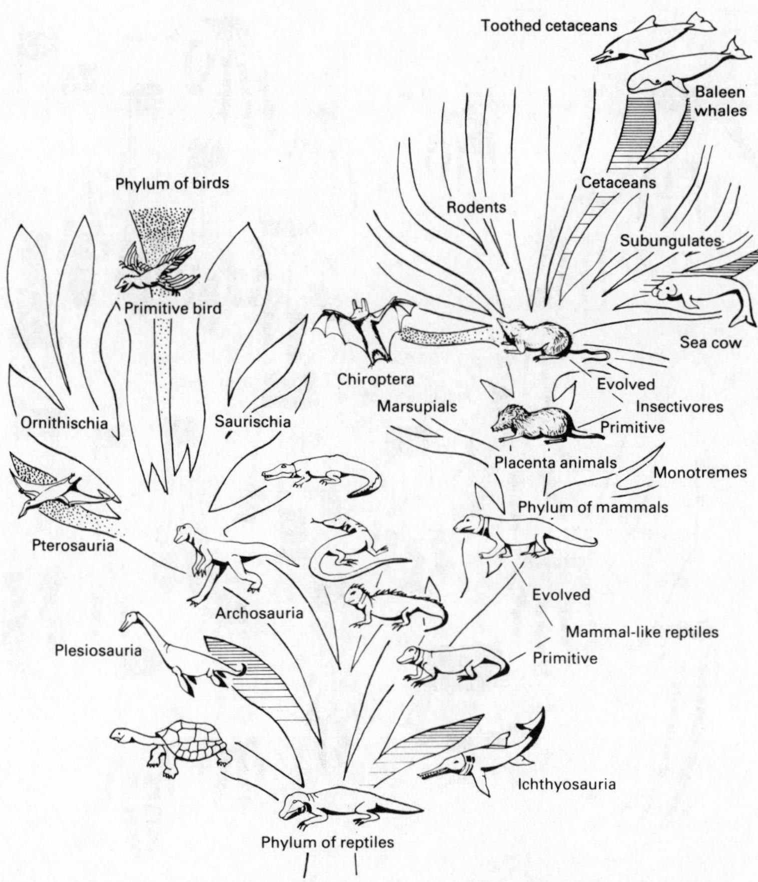

FIG. 44 A dispersed situation of convergent similarities, exemplified in adaptations to flight and water within the realms of reptiles and mammals. The field of similarities is organised according to the predominant harmonious-divergent similarities. Note here the chance distribution of pterosaurus, birds and chiroptera, as well as that of ichthyosaurus, plesiosaurus, sea cows and cetaceans. To the extent that these become more similar than the associated primitive forms, functional analogies are involved whose identical causes must therefore lie outside the organisms (combined and simplified from Romer, 1966).

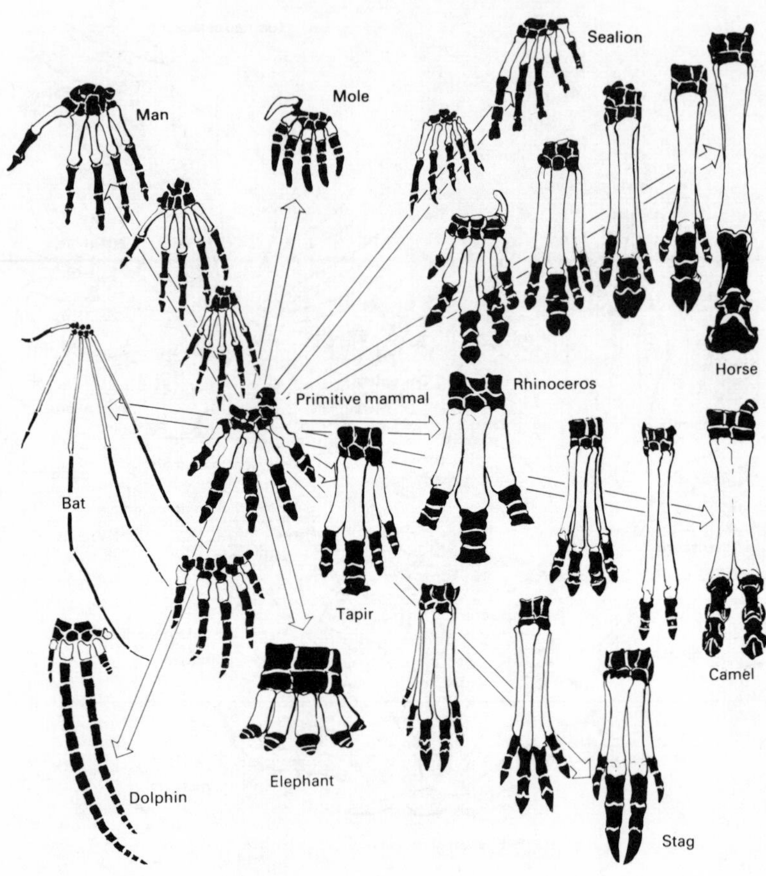

FIG. 45 A closed field of divergent similarities, showing examples of the left-hand skeleton of some recent and fossil mammals. The field is arranged in series according to the change in similarities. Although the arrangement depends on a consideration of all features available to the systematist, it can be shown that even the details portrayed follow the generally harmonious divergencies. To the extent that the basic patterns of these similarities recur in spite of functional deviations, it is a matter of homologies, the identical causes of which must now lie within the organisms (from Gregory, 1951; Thenius and Hofer, 1960; Romer, 1966).

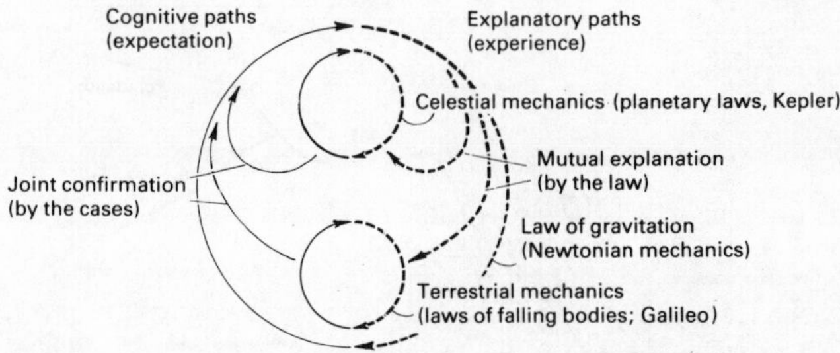

FIG. 46 The hierarchy of propositions and laws, represented by three laws of mechanics. What we regard as an explanation, proves to be the deductive relation of a higher proposition to its cases, again obtained by description (cf. Fig. 40).

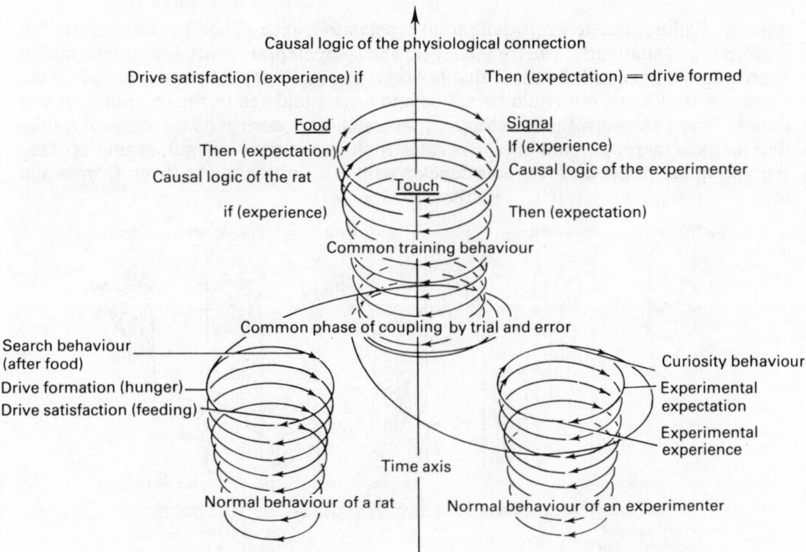

FIG. 47 Survey of a cycle of causes as exemplified by a training course. The causal relation consists equally of the normal behaviour of the experimental animal and that of the conductor of the experiment. The behaviour of the one becomes the cause for the others to continue with their behaviour. The two cycles of behaviour consist of a spiral of expectation and experience, which is as long as the two biological histories. In teaching and learning behaviour they come together through trial and error (from Riedl, 1978/79).

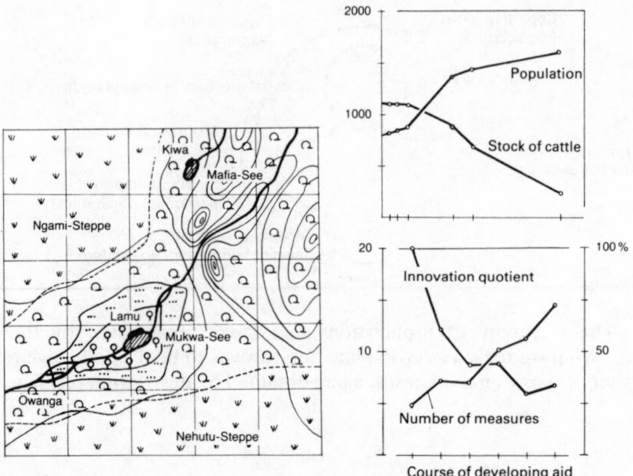

FIG. 48 Failures due to executive causal expectation, exemplified by the catastrophes in Dörner's "Tana-land". Twelve intelligent and competent students were set the task of improving the living conditions in that fictitious country (left). All the conditions of the country were known but could be varied and were simulated in the computer. It was shown, "that, almost without exception, the candidates destroyed the original stable-structure and thereby created frequent catastrophic conditions". Right, examples of the average values of the unfavourable developments and interventions (from Dörner and Reither, 1978, p.527).

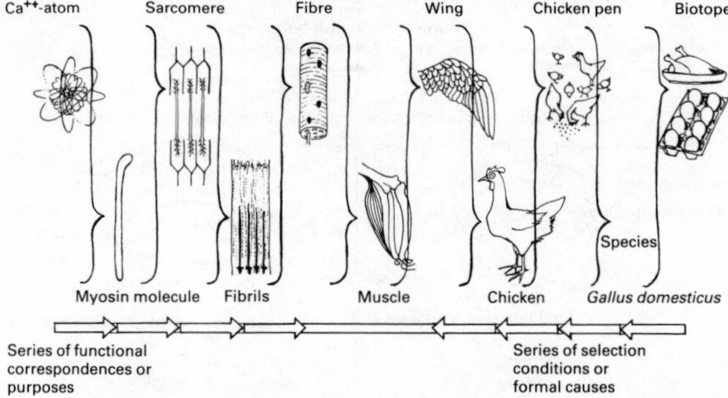

FIG. 49 From the hierarchy of purposes, a simplified series, using chickens as examples. The actual higher system (right) contains the conditions for selection of form or function for its next lower system (left), from which lower systems shown in the figure, only one is followed further. Each functionally correct lower system is regarded as purposeful and its purpose is explained from its function in the higher system; from the purpose of chickens in our kitchen right down to the purpose of the Ca^{++} atom in the "Ca-pump", to the oar stroke of the molecular cross-linkages, which is the cause of the contraction of the muscle.

FIG. 50 The planned use of tools in captivity. On the right, the capuchin monkey "Pablo" is attempting to release his feeding bowl, on the left, the chimpanzee "Julia" is opening a series of boxes which, at the time, contained a quite different tool for opening other boxes; the tools are set out above (from Döhl, 1966; Rensch, 1973).

FIG. 51 Planned use of tools in nature. Here, a suitable twig is broken off by a chimpanzee, suitably bitten into shape and, introduced into the opening, scratches into a termite nest, in order to pick out to be licked off the termites which attach themselves to the twig by biting (from photographs in van Lawick-Goodall, 1967).

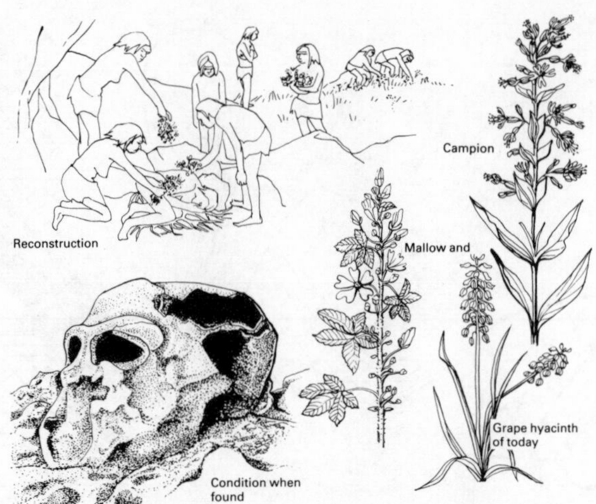

FIG. 52 Early metaphysical purposive notions exemplified by Neanderthal man. In 1960 Ralph Solecki found Neaderthal remains in the Shanidar cave (Zagros mountains, Iraq). Arlette Leroi-Gourhan (Paris) analysed the soil samples. In them were found such thick layers of mallow, campion and grape hyacinth pollen that their abundant addition to burials 60,000 years ago is beyond doubt. Were they placed there as medicinal herbs, because some are still used today in Iraq for poultices and medicines? Or were the same motives operating then which induce us now to lay flowers on a grave? (from Solecki, 1971; Constable, 1973).

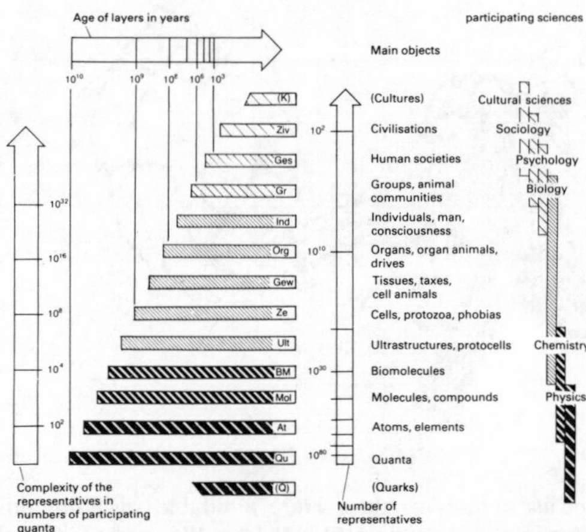

FIG. 53 The stratified structure of the real world, greatly simplified to a dozen layers with approximately equal distances in their growing organisation. The pyramids emerge from the age of the layers and the complexity as well as from the number of representatives (from Riedl, 1978/79).

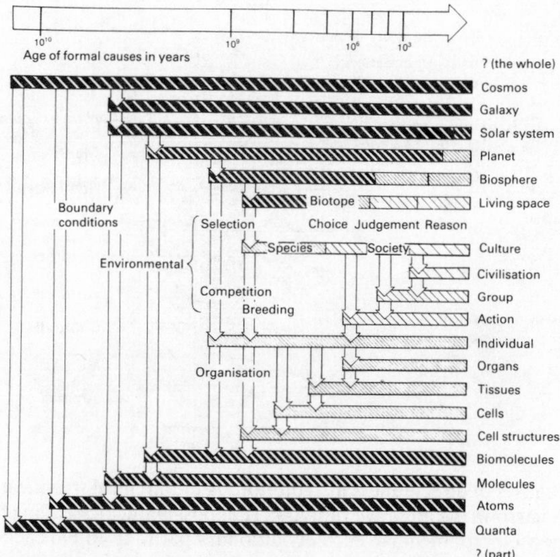

FIG. 54 The development of formal causes in relation to the evolution of the stratified structure of the real world (cf. Fig. 53); arranged according to the age and complexity of the layers. The development between the "part" and the "whole" is of interest, the step-wise differentiation of formal conditions and the change of our terms for them from the "boundary conditions" up to "reason" (from Riedl, 1978/79).

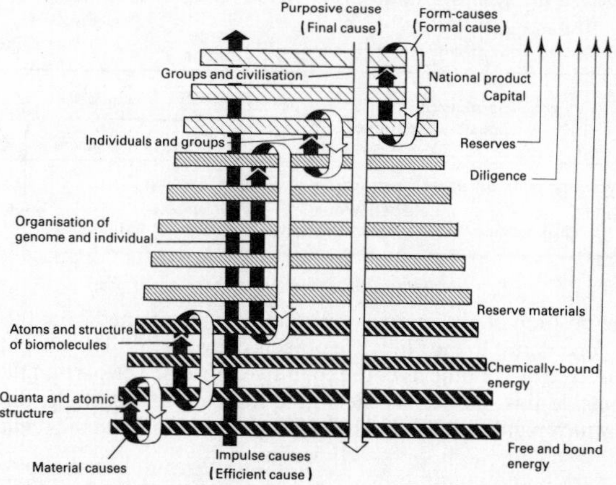

FIG. 55 The four forms of causes relative to the layer structure of an individual and of his civilisation. The cycles of material and formal causes change from layer to layer, while the final and efficient causes penetrate unchanged through the whole structure; some notes on impulse causes and the forms of energy have been inserted (from Riedl, 1978/79; compare also Figs 53 and 54).

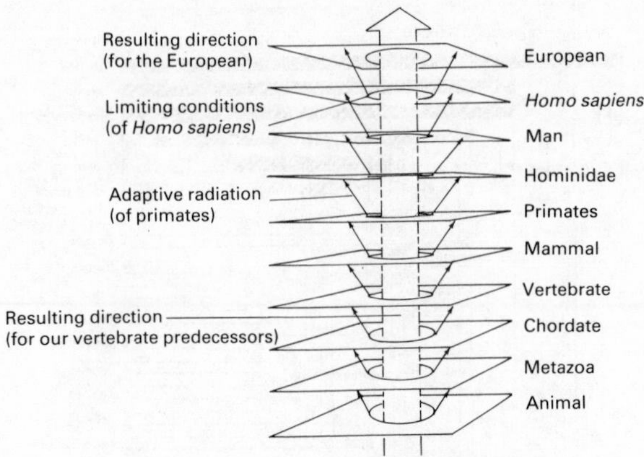

FIG. 56 The causes of directedness in evolution. A hierarchical stratification of boundary conditions narrows the adaptive radiation (the possibilities extending through adaptation to the new environment) in each evolutionary pathway and in each layer; so that our course, too, without prior intention, ultimately receives a definite goal.

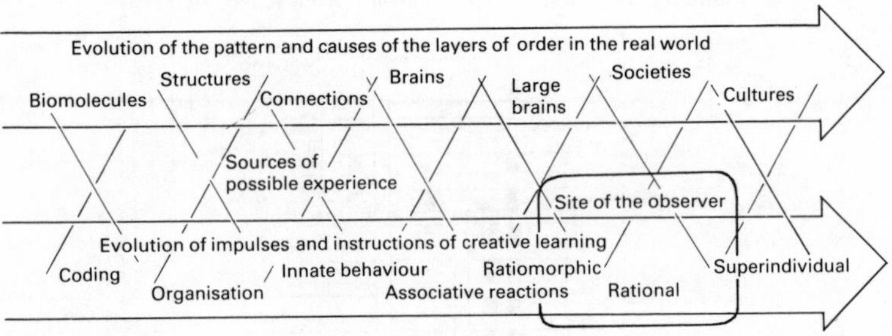

FIG. 57 The position of evolutionary epistemology, presented from the standpoint of the observer, in contrast to the objective objects of evolution. The site of the observer is only partly covered by the objects of the cognitive process. The parts of the learning product lying outside this, as well as the whole learning matrix itself, the classification layers, from which regularity is extracted through learning, lie in the realms of objective science.

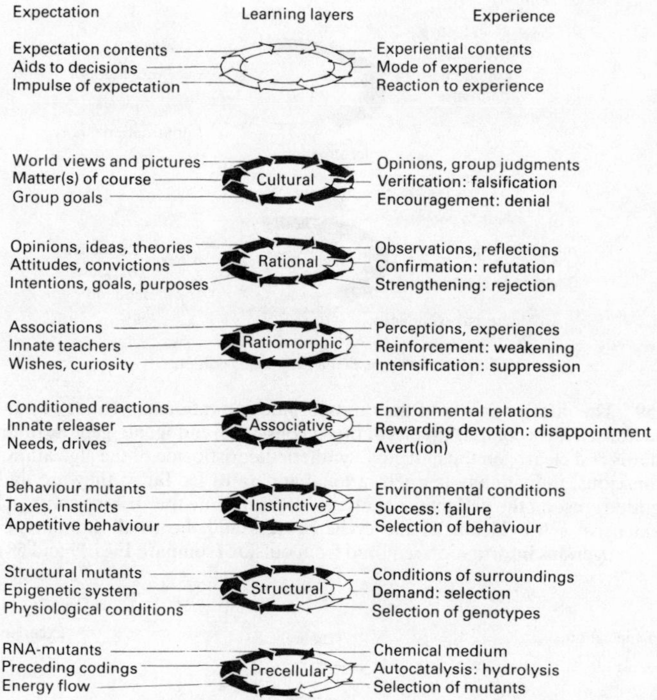

FIG. 58 The evolution of the learning algorithm. On the left, the three parts referring to expectation; on the right, those referring to experience. The two arrows in the middle axis, in each case, stand for the information available to the learning individual from outside (upper arrow) and from within. The endogenous portions (taken up into the learning system) are shown in black. Note their increase in the process of evolution (and compare Figs 29, 30 and 59).

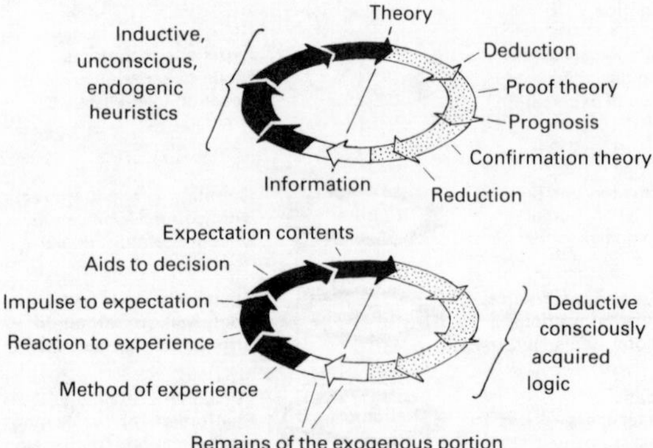

FIG. 59 The differentiation of the ratiomorphic and rational portions of our learning instructions. The accomplishments of the unconscious, endogenic-genetic ratiomorphic apparatus rest clearly on the inductive synthetic-heuristic side of the algorithm; those of the conscious, individually learned rational apparatus (so far as they are undividedly recognised) rest on the deductive, analytical-logical. Note the striking contrast between the extensive self-evidence of the cyclic process and the small amount of residual exogenous information required from outside (compare Figs 29 and 58).

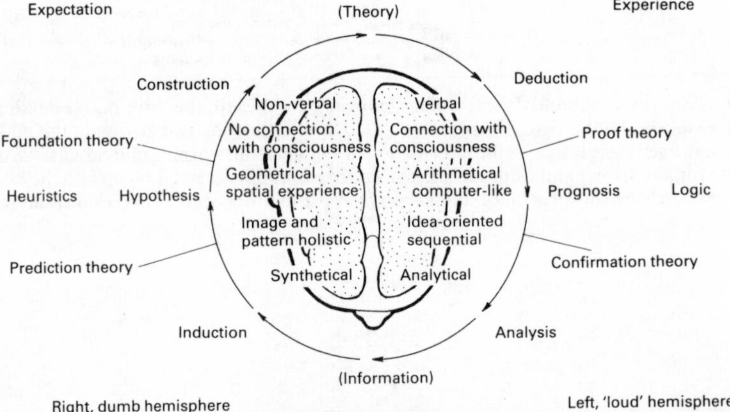

FIG. 60 The parallelism of hemispheric and cognitive functions of the human brain. The complementary functions of our brain hemispheres are inserted according to the experiences of neuropsychology (after Levi-Agresti and Sperry, 1968; from Eccles, 1975), the complementary performances of the cognitive process according to the experiences with the dynamic of scientific theory formation (Oeser, 1976 and the results of our own evolutionary investigations). Note this extensive agreement (compare, in addition, Figs 59, 58 and 29).

APPENDIX

NOTES

Introduction

1 Quoted from P. Weiss (1971; p.231).
2 On the pecularity of philosophy referred to, Brockhaus writes: "Philosophy as such is not predetermined but from time to time brings forth its notion itself. This results in a multitude of differing 'definitions' . . . in the shape of differing philosophies . . . none of them has proved to be durable in all respects."
3 Compare C. Vollmer (1975; p.183).
4 These are clearly presented particularly in the writings of Konrad Lorenz (1965, 1973) and I. Eibl-Eibesfeldt (1978).
5 For guidance on these developments, see M. Eigen and R. Winkler (1975) for the prebiotic realm, N. Chomsky (1968) for the area of speech development or E. Lenneberg (1972), I. Eibl-Eibesfeldt (1970) or O. Koenig (1970) for the cultural sphere, I. Rechenberg (1973) for technology, as well as Th. Kuhn (1967), E. Oeser (1976) or K. Popper (1959) for the development of science.
6 The "Philosophie zoologique" of Lamarck appeared in 1809, Lyell's "Principles of geology" in 1830 and Darwin's "On the Origin of Species" in 1859.
7 See glossary.
8 Here we are concerned with the epistemological induction problem (cf. glossary), to which we shall return in greater detail in the following chapters. The first real formulation of these problems was due to D. Hume (1748).
9 In P. Weiss (1971; p.231).
10 The most important works that summarise scientific experience in this area are: I. Eibl-Eibesfeldt (1978), K. Foppa (1965), E. von Holst (1969), F. Klix (1976), K. Lorenz (1965, 1973, 1978), I. Pavlov (1972) and B. Rensch (1973).
11 R. Riedl (1975).
12 In the present volume it can be shown that we rediscover Kant's a priori in innate thought patterns, which hitherto (e.g. in R. Riedel, 1975) was not so evident.
13 M. Eigen and R. Winkler (1975).
14 E. Schrödinger (1944).
15 J. Piaget (1973, 1974).
16 K. Lorenz (1973) and I. Eibl-Eibesfeldt (1978).

17 N. Chomsky (1968) and E. Lenneberg (1972).
18 O. Koenig (1970, 1975).
19 R. Riedl (1976).
20 It is known that S. Freud spoke of the "unconscious" and C. Jung of the "collective unconscious". To what extent these views coincide with those in our book would be worth investigating.
21 E. Brunswik (1955).
22 In 1941, in the work on "Kant's teaching of the a priori in the light of present day biology".
23 D. Campbell (1966).
24 K. Popper (1959).
25 E. Oeser (1976).
26 D. Campbell (1966).
27 G. Vollmer (1975).
28 In this connection the term goes back to L. von Bertalanffy (1955).
29 R. Kaspar: "Introduction to biological epistemology" (1977), published as a paper in the Hochschülerschaft der Universität Wien. See also R. Kaspar (1979).

Chapter 1

1 Quoted from K. Popper (1972) and W. Stegmüller (1971; p.13).
2 This was published in 1809. An introductory survey to the historical background is given e.g. by S. Mason (1974).
3 Consult the readable "Pleasures of philosophy" by W. Durant (1953).
4 This view of truth as an agreement between thought and reality corresponds to the theoretical truth concept, which belongs to the group of attributive concepts of truth; this can be compared with the group of substantive concepts of truth ("Essence of Truth"). See e.g. A. Diemer and I. Frenzel, ed. (1967; p.329).
5 Quoted from "Essay concerning human understanding" (1690).
6 A. Diemer and I. Frenzel, ed. (1967; pp.262-280) give an introductory survey on the philosophy of positivism and neopositivism. There is a brief description in the glossary.
7 Parmenides, born in Elea about 540 B.C., together with Zeno and Melissos, belonged to the school of Eleatics. Of his didactic poem "On Nature" 155 hexameters have been preserved. The first part of this poem ("On Truth") is the beginning of epistemology; in it he draws up a model of subject-object relationship, the theme of which remains valid.
8 In K. Lorenz (1973; p.10).
9 Thus the subject should be seen objectively, but the object subjectively.
10 This is Descartes' essential foundation for epistemology in the history of philosophy. It represents one attempt to find some indubitable beginning to epistemological considerations.

11 Compare a key work of each of these philosophers: Augustin (354-430): "De civitate Dei" in 22 books, completed in 428. F. Schelling (1775-1854): "Ideas on the Philosophy of Nature", 1797; G. Hegel (1770-1831): "Phenomenology of the Spirit", 1806.

12 Since the commentaries of Boethius (480-525), every scholastic philosopher became some kind of disciple of Aristotle. It must have been only too obvious to scholastic philosophy, whose main concern was to establish revelation on the basis of reason, to consider final causality as supreme, because the idea of a purposeful, predestined creative event could thereby be interpreted as a consequence of "natural causal connections".

13 From the type notion, conceived by J. W. von Goethe, developed the so-called idealistic morphology, which sees the type of species, genus, family etc., as a prototype similar to Platonic ideas. Present day biology sees type as a reality lying in genetic relationships. See B. Hassenstein (1951), A. Remane (1971), R. Riedl (1975) and R. Kaspar (1977).

14 In the preface to "Critique of Pure Reason".

15 With respect to epistemology, solipsism represents a radical consequence of idealism, in that the solipsist considers only his own thoughts as real, but all others as imagination. For example, M. Stirner (1861) represented solipsism in a socially orientated form.

16 Quoted according to K. Popper (1972).

17 J. Lederberg (1963), for example, represented this position in the CIBA-Foundation symposium.

18 Thus it was one of the central problems for theologists who seek to bring human freedom into harmony with divine predestination, a problem Augustin struggled with and which occupied the church fathers in the Middle Ages. Pope Innocent III (1198-1216), and Thomas Aquinas, found a solution by designating everything as sin that occurred against one's convictions. Compare, for example, R. Zorn, (1952).

19 As in the work by J. de Lamettrie "L'homme machine" (1747).

20 Actually by the discovery of uncertainty in the behaviour of subatomic elementary particles. The consequences are discussed, e.g. by W. Heisenberg (1969).

21 This may happen when an alteration in the genetic substance (a mutation) occurs by atomic chance and this change in the genome is expressed in an alteration of a characteristic.

22 Further information may be found in his work "Man in the Cosmos" (1957).

23 J. Monod, for example, writes thus: "He (man) now knows that he has his place, like a gypsy, at the margin of a universe that is deaf to his music, indifferent to his hopes, sorrows or crimes". (1970).

24 Cf. R. Riedl (1976; p.300). Quoted from W. Durant (1953; p.16).

25 This subject is discussed in D. Hume's "An enquiry concerning human understanding" (1748).

26 Chiefly in “Critique of Pure Reason” (1781) and “Critique of Judgement” (1790).
27 In B. Russell: “History of Western Philosophy” (1961). Quoted from K. Popper (1972).
28 For the philosophy of R. Carnap, see P. L. Schilpp, editor, (1963). An introductory survey of the induction problem is given by, e.g., W. Stegmüller (1971).
29 K. Popper established his point of view in “Knowledge of conjecture: my solution of the induction problem”; reprinted in K. Popper (1972).
30 Quoted from A. Diemer and I. Frenzel, ed. (1967; p.169).
31 Referred to in E. Oeser (1976).
32 Especially in “Die Rückseite des Spiegels” (1973). The theme is similarly discussed in R. Riedl (1976).
33 P. Berger and Th. Luckmann (1969) in particular, as well as P. Watzlawick (1976) have set out these facts very clearly.
34 Numerous examples for this are to be found in P. Watzlawick (1976).
35 In H. Albert (1968; p.13).
36 Quoted from K. Popper (1972).
37 For further information see F. Kluge (1967[20]).
38 Since this could mean various things today, we refer to the statement in Brockhaus’ Encyclopedia, which provides a representative cross-section.
39 The term “purpose” is to be understood here not in the philosophical-teleological sense but in its original meaning. More about this especially in Chapter 5.
40 Cf. E. Schrödinger (1944).
41 Experts can refer to M. Eigen and P. Schuster (1977); a simple introduction is given by M. Eigen and R. Winkler (1975).
42 M. Eigen (1976) has formulated this principle mathematically as the origin of information. Epistemologically, as the origin of living order, it was derived by R. Riedl (1975). A short summary can be found in R. Kaspar (1978).
43 A clarification of this (seeming) tautology, as regards the molecular region, is to be found in M. Eigen and R. Winkler (1975). Obviously such a tautology does not exist for the biologist, since he does not assess selection value or “fitness” simply by the survival of the organism.
44 Quoted from P. Weiss (1971; p.231).
45 Compare Goethe’s verse “Wär’ nicht das Auge sonnenhaft, die Sonne könnt’ es nie erblicken” (“If the eye were not radiant, it could never behold the sun”), which almost literally goes back to Plotinus (3rd century A.D.), the founder of the Neoplatonic school.
46 For the background of this thought, see I. Fetscher: “Philosophy of history”, A. Diemer & I. Frenzel, ed. (1967; p.84). I. Kant wrote about it in his work “Idea for a universal history” (1784).
47 For example, in the well-known work “The Social Contract”, 1762.
48 Quoted from I. Kant (1784).

49 Compare E. Schrödinger (1944).
50 Refer to K. Lorenz, in P. Weiss (1971).
51 See K. Lorenz (1973).
52 See glossary.
53 See glossary.
54 If the average life span of a species is put at 10^6 years. Compare E. Mayr (1967).
55 To such capacities for error we shall return in the last sections of Chapters 2, 3, 4 and 5.
56 K. Lorenz (1973), in particular, has described these conditions in detail.
57 In this connection, compare P. Feyerabend (1970), Th. Kuhn (1967), E. Oeser (1976) and K. Popper (1959, 1972).
58 This is described in detail by I. Eibl-Eibesfeldt (1978) and K. Lorenz (1963).
59 Quoted from the course given by K. Lorenz at the University of Vienna in the winter term of 1976.
60 From F. Dessauer (1958).
61 "Candide ou l'optimisme" by J. Voltaire (1759) was considered a mockery of G. Leibniz's "Essais de théodicée sur la bonté de dieu, la liberté de l'homme et l'origine du mal" (1710).
62 For guidance see e.g., H. Hermes (1961); for algorithm, see glossary.
63 This concept of order as well as the methods for approaching the ordering phenomenon from the epistemological point of view, are developed in R. Riedl (1975, especially Chapter 1); summarised in R. Kaspar (1978).
64 From B. Russell (1948[20]).
65 This is about some arguments that support the so-called reality postulate; cf. G. Vollmer (1975; pp.35-39).
66 The term "hypothetical realism" was first used by D. Campbell (1959) and K. Lorenz (1959).
67 Quoted from G. Vollmer (1975; p.35)
68 The positions of different forms of realism have been clearly set out by G. Vollmer (1975).
69 For details, refer to E. Brun (1912).
70 The philosophical thesis "Natura non facit saltus", as an expression of the continuity principle, dates back to J. Fournier (1613) and was later taken over by Leibniz, Linné, Goethe and Schopenhauer.
71 In his book "What is life" (1944).
72 For the specialist it is most recently outlined in M. Eigen and P. Schuster (1977); the theory of the hypercycle is described for the general reader by M. Eigen and R. Winkler (1975) as well as by P. Schuster (1972).
73 A well-known example of an unconditioned direct reflex is the so-called patellar tendon reflex. A sudden pull (or pressure) on the kneecap ligament sets off a spasm of the four-headed thigh muscle. The reflex is used clinically for testing hip marrow. Its biological significance lies in the rapid

accommodation of the contractions of the leg muscles to movements on walking.

74 See R. Riedl (1976).

75 This attitude was taken by the American school of behaviourism, e.g. by J. B. Watson (1925) or B. Skinner (1973). Not only did it offer no explanation but it is plainly false.

76 If, in the case of the dog, one assumes only 16 connection points between the inner ear, brain and the muscles of the salivary gland, this yields 16! (16 factorial) possible permutations, that is, roughly 2 x 10^{13} (20 million million). The whole life span of a dog (about 3 x 10^{8} sec) is too short for even one single association to be correctly hit by chance in this way.

77 See K. Lorenz (1973).

78 In the book of the same name (1976), H. von Ditfurth has set out the natural history of consciousness.

79 G. Vollmer (1975; p.55) has assembled from the history of philosophy different epistemological explanations of our ideas on the three-dimensional nature of space.

80 A survey of optical illusions can be found in e.g. R. Gregory (1972).

81 For this, compare R. Riedl (1976).

82 First in E. Brunswik (1955).

83 The most important publications on this are N. Chomsky (1968) and E. Lenneberg (1972).

84 A. Gehlen, especially, has often pointed to this circumstance, as in the work of 1940.

85 See K. Lorenz (1959).

86 Quoted from K. Popper (1972).

87 Quoted from K. Popper (1972).

Chapter 2

1 Quoted after K. Popper (1972). The quotation from G. Leibniz refers to the formulation by J. Locke (1690): "Nothing is in the understanding that was not previously in the senses". Leibniz gave this answer in his work of 1704.

2 "Truth and lies" is one of the oldest fables which were handed down in numerous settings: Chinese, Tibetan or Hebrew, since the 12th century in Europe. Compare A. Wesselski (1947).

3 A popular account of the Gilgamesh epic will be found in C. Ceram (1949) or H. Schmökel (1966).

4 We recall the panorama of philosophical opinions and attitudes from chapter 1.

5 In "Faust, part I", line 364. Socrates says in his "Apology" "By this small amount I seem to be wiser in that I do not even believe that I know what I do not know" (see Plato, Apology; 21d).

6 See glossary ("idealism").
7 Just as solipsism refutes itself as soon as it is held. For are not the men whom the solipsist wishes to convince of his views products of his imagination?
8 In W. James (1907). See also W. Corti, ed. (1976).
9 This is the problem of justifying inductive inference. More about this later.
10 A. Einstein's famous answer to the probabilistic interpretation of subatomic phenomena was: "surely you cannot believe that God gambles!" (see A. Einstein, M. Born; 1969).
11 See I. Kant (1781).
12 K. Popper (1959) uses this example.
13 See R. Riedl (1976).
14 Aristotelian logic is found in the "Organon"; see also G. Frege (1879).
15 Here it is a matter of the Lac-Operon. Further details can be found in the text-book by C. Bresch & R. Hausmann (1972) and also in J. Monod (1959).
16 K. Lorenz, in particular, has described it clearly (1973; p.67).
17 Some initial idea of the time scales of evolution can be obtained if one considers the preservation times of innate characteristics; on average, this amounts to 10^5 to 10^8 years, even longer in the case of "living fossils"; compare R. Riedl (1975; p.168 ff.).
18 The details of this process are described by B. Hassenstein (1973) and K. Lorenz (1973, 1978).
19 See I. Pavlov (1927). Pavlov's conditioned behaviour is no simple conditioned reflex but a conditioned appetitive behaviour, because the dog reacts with his whole food begging behaviour insofar as he is not firmly bound for experimental reasons. The conditioned reflex (salivary secretion) is only one component of this behaviour.
20 This term is derived from N. Hartmann (1964). It defines the hierarchical classification of the complexity layers of the real world from quanta, through biostructures up to the highest categories. Causal connections act both on the more complex as well as on the simpler layers (R. Riedl; 1976, 1978/79). A higher category, in its entirety, is always something qualitatively different from the sum of its elements, since with every new level of integration new system laws arise.
21 In K. Lorenz (1943, 1973).
22 For orientation, refer to K. Lorenz (1973; pp.106-113). The notion of "imprinting" is explained in the glossary.
23 For this, compare the details in K. Lorenz (1973) or in I. Eibl-Eibesfeldt (1978). An extensive presentation of the imprinting phenomenon is given by E. Hess (1975).
24 For the concept of the algorithm (see glossary) compare for example, H. Hermes (1961).
25 Cf. the heuristic notion of epistemology (see glossary).
26 Quoted from I. Kant (1783).
27 In W. Wickler (1968). See also glossary ("mimicry").

28 In S. Vogel (1978).
29 See P. Kuyten (1962).
30 In D. Grant and L. Schipper (1952).
31 In K. Foppa (1965).
32 This was first described by W. Craig (1918). An introductory survey is given by I. Eibl-Eibesfeldt (1978).
33 Quoted from B. Hassenstein (1973; p.207, 209, 210). On p.207 he gives a cybernetic switch scheme of conditioned reactions.
34 In P. Watzlawick (1976).
35 For the oldest vertebrate animals belong to the silurian period (about 5 x 10^8 years ago).
36 In E. Brunswik (1955). See also glossary.
37 Mainly because it has not yet been possible to define all variables including their natural interactions.
38 On the age and time-dating of the development of man, refer to M. Edey (1973).
39 See A. Koestler (1966).
40 This creative aspect in scientific research becomes clear, for example, when a solution of a complex problem suddenly "dawns" on us (notice the delicacy with which language designates this process); at first we do not know rationally how the solution will actually appear.
41 The role of form perception in these processes has been clearly stressed by K. Lorenz (1959).
42 The psychoanalytical investigations of S. Freud and C. Jung have indeed surmised as much.
43 See D. Campbell (1966).
44 This expression is to be understood in the sense that the perceiving subject really represents a "mirror" in which the contents of awareness and thoughts are reflected. The central nervous mechanism which makes this image possible, is then the back side of the "mirror". See K. Lorenz (1973) and R. Kaspar (1979).
45 This special problem-solving process is explained in more detail in the glossary.
46 Further information on this subject will be found especially in R. Riedl (1975, 1976).
47 These axioms (see A. Kolmogorow, 1933) read: 1. To each event A is assigned its probability, a number P(A), whence 0 is less than or equal to P(A) which is less than or equal to 1; 2. P(E) = 1, i.e. the probability of the certain event is 1; 3. If the events A_1, A_2 ... A_n are mutually exclusive, then the probability that either A_1, or A_2, or ... A_n occurs, is equal to the sum of the individual probabilities:

$$P(A_1 \cup A_2 ... \cup A_n) = \sum_{i=1}^{i=n} P(A_i)$$

48 See F. Ramsey (1931) and B. Finetti (1970).
49 Quoted from F. von Kutschera (1972; vol.1, p.46).
50 Quoted from F. von Kutschera (1972; vol.1, p.47).

51 In E. Oeser (1976); especially volume 3, p.119.
52 The details are given in I. Kant (1781).
53 This is established in detail in "The Critique of Pure Reason".
54 In I. Kant (1781).
55 For example, in K. Lorenz (1941, 1973) and D. Campbell (1974).
56 A theoretical basis for this statement is given by K. Popper (1959, 1972).
57 In E. Brunswik (1934), K. Lorenz (1973), G. Vollmer (1975).
58 See T. Bayes (1908). Compare also the survey in E. Oeser (1976; vol.1, p.55 ff).
59 In R. Riedl (1975).
60 The details of the Weber-Fechner law can be looked up in every text-book of psychology or physiology of sensation. Some idea may be obtained in Hubert Rohracher (1971).
61 See T. Stoppard (1967).
62 This experiment was performed many times with about 150 students. It was agreed that the conductor of the experiment would settle for "heads" but the whole group for "tails"; two coins stuck together were used, which showed "heads" on every throw. The swindle (thus the absence of the expected chance distribution) was often recognised on the fourth or fifth attempt. (One should not be deceived by the apparently trivial nature of such experiments, since our first impression that this should be obvious at once shows how deep the expectation of certain probabilities is rooted in our minds. These expectations are thus not self-evident.)
63 See F. Ramsey (1931) and B. de Finetti (1970). A summary review is given by F. von Kutschera (1972; vol.1).
64 For logic, G. Frege coined the term "truth value", which referred only to the circumstance of whether a proposition or statement is true or false within a certain language system. The logical truth of a proposition relates to the correctness of its formal structure within a given calculus. Logically true propositions can be false even in content (factually), like, say, the sentence: If some insects are vertebrates, then some vertebrates are insects.
65 This relativising of the number of real deceptions refers to the fact that, in the case of chance events, a relative number W of cases is correctly guessed simply by chance. In the case of a repertoire of two, this is indispensable, but for a repertoire from three or four it can be practically neglected.
66 To this more exact consideration of the ratiomorphic probability calculus, Günter Wagner has made a decisive contribution.
67 In the case of a repertoire of two this is particularly simple. With a repertoire of more than two there are several possibilities of continuing the series non-periodically. Of the possible events, we must then choose the one that so far has occurred least often.
68 In A. Remane (1971). By "essential similarity", in biology, we mean the homology (see glossary) of structures.
69 Compare G. Frege (1879) and B. Russell and A. Whitehead (1910/13). An introductory review is given in A. Diemer and I. Frenzel, ed. (1967).

70 Compare the summaries in R. Carnap (1976) or K. Popper (1959); details in the context of the induction problem are given by W. Stegmüller (1971).

71 Quoted from K. Foppa (1965, p.19); compare also with L. Pickenhain (1959).

72 See E. Brunswik (1934), L. Humphreys (1939) and the survey in K. Foppa (1975).

73 In D. Grant, H. Hake and I. Hornseth (1951).

74 In this case, it is a question of fortuitous or determined series of, in each case, 2-4 possible events. How great the experience must become in order clearly to recognise chance or intention, should be tested. The experiments were performed by Claudia Rohracher in the course of a laboratory study in my institute.

75 Compare H. Simon and K. Kotovsky (1963). A review of the psychological tests is given by R. Brickenkamp (1975), for the psychology of learning see K. Foppa (1975) and K. Joerger (1976) as well as W. Krause (1970) and F. Klix (1973).

76 See E. Oeser (1976; vol.3, p.118, also the schema on p.119).

77 In F. Ramsey (1931), B. de Finetti (1970) and L. Savage (1967).

78 Precisely the converse definition of subjective probability is given by F. von Kutschera (1972; vol.1,p.47).

79 Here, as shown in our experiment in the lecture theatre, these reasons can be related to tradition, social position, to the university, to the person conducting the experiment or to simple emotions.

80 See K. Popper (1959), I. Hacking (1965). C. Peirce, collected by C. Hartshorne and P. Weiss.

81 Quoted from F. von Kutschera (1972; vol.1, p.123) according to R. Carnap (1962).

82 See H. Simon and K. Kotovsky (1963) and W. Krause (1970).

83 The increase in certainty lies between the corrected and the uncorrected calculated values. Thus the effect of the confirmation is calculated, but the confirmation on the basis of chance (e'-W) in place of (e') is insufficiently considered. Therefore the certainty found in the case of regularity is overestimated, but that in the case of chance underestimated.

84 See R. Riedl (1976) or in greater detail A. Bavelas (1957).

85 See, for example, K. Popper (1972).

86 Thus, the formulation of order as regularity times its application; see R. Riedl (1975, 1976).

87 As an example, D. Hume (1748) has written: "I venture to put the proposition, as generally valid and tolerating no exception, that the knowledge of this connection (of cause and effect) is in no case obtained by an act of thought a priori but is derived exclusively from experience." Since, therefore, according to Hume there is no logical pathway leading from the observed to the non-observed, there can be no correct conclusion which contains more truth than its premisses. That is the problem of induction (see glossary).

88 Quoted from W. Stegmüller (1971; p.13).
89 This number is obtained from 2 million known species plus about 500,000 systematic categories, multiplied by the average (20) of its diagnostic characteristics.
90 In K. Popper (1959).
91 Illustrations of the black swan (Cygnus atratus) and the black-necked swan (Cygnus melanocorphyrus) may be found in B. Grzimek (1968; vol.7) as well as in Fig. 14.
92 Quoted from W. Stegmüller (1971; pp.16 and 17).
93 Quoted from E. Oeser (1976, vol.3, p.68).
94 In E. Oeser (1976; vol.3, p.71ff), where the interested reader will find further references.
95 The swans belong to the family of the Anatidae (related to the goose) and form the subfamily, Cygninae (swans).
96 The following quotations are a reversal of W. Stegmüller (1971; p.19). They deal with an allusion to a rule of foresight by M. Black (1954).
97 One may reckon that about two million species mainly in the realm of the lower organisms are still to be discovered (see R. Riedl, 1970).
98 See B. Hassenstein (1973), I. Eibl-Eibesfeldt (1978) and K. Lorenz (1973, 1978).
99 Quoted from W. Stegmüller (1971; p.17).
100 By the actuality principle we mean operating with the assumption that present regularities in nature are, in principle, the same as those in the past. For the application of this principle, see I. Kant (1755), P. Laplace (1796), J. de Lamarck (1809), C. Lyell (1830) and C. Darwin (1859).
101 See R. Riedl (1975, 1976).
102 In K. Lorenz (1943, 1973), E. von Holst (1969) and N. Tinbergen (1951).
103 See E. Oeser (1976; vol.3).
104 Quoted from C. Hempel (1945).
105 Beside the priceless poems of C. Morgenstern, H. Bosch, in particular, in many of his pictures has presented the resolution of natural interdependencies in "organisms", which for example are composed of features of reptiles, worms, butterflies, birds and mammals. Even here it is shown that a complete resolution of interdependencies cannot be imagined (see also R. Riedl; 1975, p.222 ff.).
106 In E. Oeser (1976, vol.3, p.75). "Ars iudicandi" means the "art of judgment", "ars inveniendi" the art of discovery of new knowledge. This latter is "ordine naturae certe prior", that is, "surely prior in the order of nature"; for "syllogistics" and "topic", see glossary.
107 The historical background and connections are described by E. Oeser (1976; vol.3).
108 Cf. P. Hofstätter (1972), T. Herrmann et al. (1977), A. Diemer and I. Frenzel (1967) and F. von Kutschera (1972).
109 A. Marfeld (1973) gives a detailed review on this.
110 Very clear models of the mechanisms of such processing may be found in B.

Hassenstein (1965).

111 See Hubert Rohracher (1965; p.7), and in a wider context R. Riedl (1976, p.235).

112 For "social construction of reality", P. Berger and T. Luckmann (1969).

113 The phenomenon of tradition in civilisation from a biological point of view has been described particularly by K. Lorenz (1973) and O. Koenig (1970, 1975).

114 See T. Kuhn (1962).

115 This is described in detail in R. Riedl (1976; p.205ff.).

116 See, for example, E. de Bono (1975).

117 For the solution of this problem, one must go beyond the square that the figure suggests.

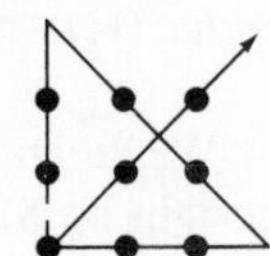

118 See M. Eigen and R. Winkler (1975) and R. Riedl (1976) respectively.

119 For fulguration, see glossary.

120 In M. Eigen & R. Winkler (1973/74, 1975).

121 See glossary ("similarity field").

122 See glossary ("homology").

123 Homoiologous similarities would be the fish form and fin structure in the shark, swordfish, a primitive amphibian, ichthyosaurus and dolphin, or the arthropod skeleton in spiders and crabs, the crista sagittalis (sagittal crest) in the gorilla and hyena. It is here a question of analogous formations on the basis of homologous structures.

124 See R. Riedl (1976) or K. Lorenz (1973).

125 According to W. Stegmüller (1971).

126 See F. von Kutschera (1972), also for further relevant references.

127 An explanation of these different probability concepts is given, for example, by R. Carnap (1967^2), F. von Kutschera (1972), E. Oeser (1976) and W. Stegmüller (1973).

128 Compare R. Carnap (1967^2).

129 H. Störig (1972) gives an introduction to astrophysics.

130 In B. Bavink (1930; p.189). Such a calculation, which Perrin made for a brick, gives a waiting time of $[10^{10}]^{10}$ years for such an event, (a number with ten thousand million digits!).

131 The mathematical basis of this state of affairs is to be found in R. Sexl (1979).

132 See E. Oeser (1976; vol.3).

133 In R. Carnap (1952).

134 Compare K. Popper (1959). The method of "quasi-induction" suggested by Popper does not solve the problem, since it consists only of deductive falsification experiments and the question, how can one arrive at a new

hypothesis, cannot be answered.

135 See E. Mach (1905), W. Whewell (1860) and E. Oeser (1976).
136 In G. Vollmer (1975; p.126).
137 In I. Kant (1781).
138 See K. Lorenz (1941, 1943), D. Campbell (1959) and G. Vollmer (1975).
139 Compare R. Riedl (1975), Chapter 1, and (1976), Chapter 3.
140 In I. Kant (1781).
141 See K. Lorenz (1973; p.79).
142 The Berlese funnel has been used by soil biologists for a long time as both a simple and effective collecting device.
143 Some cases of this kind have been mentioned, for example, by H. von Ditfurth (1976); thus it is highly significant that energy dispensers taste sweet and pleasant, but dangerous when the same taste sensation is released by lead acetate or other poisons.
144 See I. Eibl-Eibesfeldt (1978).
145 In Hubert Rohracher (1965).
146 Some information on early human history is given by G. Constable (1973).
147 K. Lorenz has often emphasised this.
148 One can think, for example, of Kepler's service for Wallenstein (see E. Oeser, 1971).
149 Chiromancy going back to the 15th century (1448, J. Hartlieb) has been applied, as chirology, since the 19th century and today, in the USA, is a widespread profession (palmists).

Chapter 3

1 Quoted from J. W. von Goethe (1790), according to the Weimar edition II, 13, p.212. In "Morphological writings" Goethe was the first to recognise the decisive accomplishment of comparison, namely, the recognition of essential similarities, for the phenomena of type and metamorphosis. The second quotation is from N. Chomsky (1968; p.82).
2 A specialist account appears in H. Schwabl (1958).
3 See also the discussion on nominalism and idealism (see glossary).
4 For rationalism, empiricims, see glossary.
5 In other languages too, the word is formed similarly, for example, in Latin com-parare = struggle together, make ready equally . In Greek the adverb "likewise" (ὁμοίως) is converted directly into the verb, namely, ὁμοιόω = to make like, to compare.
6 The origin of this idea lies in the so-called principle of indiscernibles of Leibniz, according to which objects are identical only if completely indistinguishable.
7 Among essential characteristics of living things as an open system may be included that, for maintenance of the form, the elements are constantly being exchanged. So, for instance, all the blood cells are replaced by new

ones in the course of about three months.

8 This ancient doctrine of eternal movement is especially expressed in Heraclitus, to whom the phrase πάντα ῥεῖ (everything flows) is attributed.

9 This was the title of an address by B. Hassenstein in which the problems of mapping concepts was discussed. See B. Hassenstein (1954).

10 Thus, one can draw boundaries within the sediments, by defining an average grain size: sand, with 0.02-2 mm diam., silt with 0.02-0.002 mm diam., and clay with less than 0.002 mm diam.

11 B. Hassenstein (1954) has pointed this out.

12 These phenomena are known from the psychology of perception and memory; they are surveyed, for example, in K. Foppa (1975) and F. Klix (1976).

13 In K. Lorenz (1959; p.131).

14 For details on this subject refer to R. Riedl (1976; especially in Chapter 8).

15 So for example, one finds claws in about 50% of mammals and tusks in only about 0.1% (in elephants and narwhals).

16 It would have to be an amazing chance if, immediately beneath a waiting tick, a wild boar had stirred up a stone which had been warmed to exactly 37°C by the sun.

17 W. Schleidt (1962) has described the historical development of this concept (see glossary). A review and further references in I. Eibl-Eibesfeldt (1978).

18 Details of this are described by I. Eibl-Eibesfeldt (1978).

19 N. Tinbergen and I. Eibl-Eibesfeldt have done this. The experimental results are described by I. Eibl-Eibesfeldt (1978).

20 The first investigations were carried out by D. Lack (1943) on the robin and by N. Tinbergen (1963) on the herring gull.

21 For a review, refer again to I. Eibl-Eibesfeldt (1978).

22 The physiology of this perception capacity has been elucidated particularly by E. von Holst (1969).

23 In K. Lorenz (1973, p.80ff.).

24 In accordance with the change in characteristic in the morphological region: see E. Mayr (1967).

25 This is evident in all modes of behaviour that have not yet adapted to new surroundings. It becomes especially clear in man himself, where the cultural and social evolution of development runs ahead of his innate modes of reaction.

26 Whilst, in fact, a new cultural creation can become widespread almost at once, what determines molecular learning are the mutation rate (10^{-4} to 10^{-6}), success probability (about 10^{-2}) and the generation series.

27 Compare, for example, K. Foppa (1965), L. Pickenhain (1959), G. Razran (1930) or W. Thorpe (1963).

28 Thus, up to 100 repetitions ("reinforcements") may be necessary for the formation of a conditioned reflex (see glossary). See P. Hofstätter (1972)

and for the dependence of conditioning on the experimental conditions K. Foppa (1965; p.35).

29 P. Hofstätter (1972) presents a review of the psychology of memory and forgetfulness.

30 Further references in P. Hofstätter (1972).

31 Even after two days of intensive stress under a dominating fellow creature of the same species, with Tupaias for example, distinct pathological changes in the kidneys can be detected; see D. von Holst (1969). For the many researches on stress in humans, refer to P. Bourne (1969) or H. Selye (1957).

32 Summarised in F. Klix (1976, p.370).

33 A general schema of the learning process is given by F. Klix (1976; p.352, especially Chapter 6).

34 This has to do with a mutual optimisation process, which D. Campbell (1966) denoted as "pattern matching".

35 In K. Lorenz (1973; p.159).

36 In B. Hassenstein (1965; p.108).

37 These accomplishments depend on the reafference principle (see glossary), which was first proposed by E. von Holst and H. Mittelstaedt (1950). An introductory review is given by B. Hassenstein (1965).

38 See K. Lorenz (1954).

39 Presented in G. Baerends, K. Brill and P. Bult (1965).

40 Particularly clarified by K. Lorenz. Compare also the works of Gestalt psychology, for example, W. Köhler (1971).

41 Extensive material on this is available in "The genetic cognition theory" by J. Piaget (1973, 1974). An introduction to this field is given by H. Furth (1972).

42 For no less a person than Max Planck used to say, "What do I care about my silly talk of yesterday!" Although one is usually somehow concerned, the daily abandonment of some favourite hypothesis or other, as Konrad Lorenz says, remains a healthy morning exercise.

43 On the contrary, one should consider that the cosmos has been in existence "only" since about 10^{17} seconds; - hence, in our case, 10^{14} attempts per second would be necessary.

44 Biological systematics operate here with the so-called auxiliary criteria of homology, which A. Remane (1971) has established. R. Riedl (1975) formulated the logical connections within morphology.

45 In K. Lorenz (1973; p.162).

46 K. Lorenz has recounted this observation in our seminar in Altenberg.

47 The investigation goes back to A. Bavelas (1957).

48 Cf. R. Riedl (1975; especially pp.53 and 54).

49 A large mammal, for example, possesses 10^7 identical hairs, 10^{13} red blood cells, 10^{16} chromosomes (in all cells) or 10^{27} nitrogen atoms. This phenomenon of redundancy (see glossary) of natural order is treated extensively in R. Riedl (1975).

50 If each individual characteristic has a chance of 1/2 (i.e. there is only one alternative), then the probability of obtaining by chance a structure with ten characters, $(1/2)^{10}$ is 1/1024. The chance of recovering this structure by chance in ten species, then amounts to $[(1/2)^{10}]^{10}$ i.e. about $1/1.3 \times 10^{-30}$. Its homology is therefore as good as absolutely certain. These considerations hold for divergent, continuous similarity fields (see glossary).

51 This knowledge of homologous structures depends on the comparison of similarities in harmonious fields (see glossary). Its explanation due to identical origin follows in retrospect. Morphology has used the homology concept since the time of Goethe. The closer relationships and their logical structure are explained in A. Remane (1971) and R. Riedl (1975).

52 An example of that is provided by the attempts of "numerical taxonomy" (R. Sokal and P. Sneath, 1963) to relinquish the homology concept and the assessing of characteristics. The discussion on the reality of the natural system (see glossary) is connected with these problems. One should compare it with the solution of this controversy in R. Riedl (1975), and with reference to the type problem in R. Kaspar (1977).

53 In C. F. von Weizsäcker (1971; p.361).

54 In biological structure research, this corresponds to the structural and positional criterion of homology, outlined in A. Remane (1971) and R. Riedl (1975; p.60).

55 Incidentally, differentiations of this kind often emerge from a similarity of elements. That makes strongly hierarchically structured systems always clear, perhaps as in the career of the recruit to the rank of general.

56 For example, in the elements, the halogens (group 7), in the case of molecules, the acids or haemoglobins.

57 A review study on the "biology of concept formation" is currently in preparation (R. Riedl and R. Kaspar).

58 In biological structure research, it is a question of the transition criterion of homology (R. Riedl, 1975; p.60)

59 In this particular, too, structure research has to be allowed for in the anti-coincidence criteria of homology; see A. Remane (1971) and R. Riedl (1975).

60 The historical connections can be followed in J. Hemleben (1969).

61 Note 57, above.

62 The propositions for formal logic may also be placed into this series. However, since Frege, this has retreated to the deductive realm of thought and inference, hence to a theory of proof which has nothing to do with the finding from which it was deduced. In the heuristic range of induction and prediction theory, it is better to speak of definitions and laws.

63 The first works on this were published by O. Koehler (1941) and J. Piaget (1946); see also J. Piaget (1974) or the surveys by B. Rensch (1973) and H. Furth (1972).

64 In J. Huxley (1929), K. Lorenz (1943, 1965, 1973) and B. Rensch (1973).

65 There is a survey in Hubert Rohracher (1971, p.349ff.), T. Hermann et al., (1977, p.92ff.); where further references are given.

66 Quoted from P. Hofstätter (1972; pp.89 and 92).

67 Quoted from K. Lorenz (1943; p.343), N. Chomsky (1968). G. Vollmer (1975) gives a survey.

68 On Gestalt psychology, see C. von Ehrenfels (1890), M. Wertheimer (1925), K. Koffka (1950) or A. Wellek (1953).

69 F. Klix (1976; p.283) must have noticed similar things, when he took into consideration "involvement with phylogenetic pre-formed stimulus treatment processes".

70 Compare also the table of categories in the "Critique of Pure Reason", B 106.

71 Quoted from F. Klix (1976). The original work was by C. Hovland (1952) and C. Hovland and W. Weiss (1953).

72 Compare in addition D. Dörner (1967), K. Goede and F. Klix (1972) and E. Hunt (1962). The quotation is from F. Klix (1976).

73 In R. Carnap's theory of induction, three conceptual steps play a special role: the classificatory notion, the comparative idea and the theoretical notion, but in all cases it is a question of deductive forms.

74 In E. Oeser (1976, vol.3; p.118).

75 In 1748 David Hume drew attention to the problems of inductive inference and Gottlob Frege, in 1879, excluded induction from the methods of logic.

76 As early as 1959, K. Lorenz recognised the difference between those who think mainly analytically and those who think synthetically and experience form. He expected that their mutual mistrust was attributable to being unable to receive the same experiences. With the discovery of the different functions of the brain hemispheres, neurophysiology confirmed this assumption. See J. Eccles (1975) and the expositions in Chapter 6 of the present volume.

77 "Esoteric", according to the more recent use of the word, would indicate "secretive". It meant, however, in contrast to exoteric, "established in the system". We now call (R. Riedl, 1975) the cause of the type "system-immanent". Cf. R. Kaspar (1977) and the criticism in B. Hassenstein (1951, 1958).

78 In this connection we are dealing with the morphological type (as distinct from the systematic).

79 This whole region of biological structure research is documented extensively in the literature from Goethe to Remane.

80 This reproach was raised by "numerical taxonomy" (R. Sokal and P. Sneath, 1963).

81 Here the well-known fable of the millipede and the spider is intended. The spider, who enviously admired the astonishing harmony in the movement of the legs of the millipede, enquired how it was that he always stretched the 161st leg when he raised the 162nd from the ground, and so on. The millipede stopped and tried to describe the course of the movement, but

nevertheless at once got into a hopeless muddle with his many legs and then could no longer progress at all, which greatly delighted the spider. The fable nicely illustrates our own experience in that the attempt to rationalise a process that goes on unconsciously, usually leads to it being stopped. A sure means of falling off a bicycle is to pay strenuous attention to trying not to fall off.

82 Compare J. W. von Goethe (1790), A. Remane (1971) as well as the technical details in R. Riedl (1975, 1976).

83 The criteria of position and structure (A. Remane; 1971) provide a common positional structure criterion of homology (R. Riedl, 1975, p.69ff.).

84 See R. Riedl (1975).

85 See E. Lenneberg (1972).

86 Thus the contents of the concept "mammal" has significance only in the context of the notion "vertebrate"; the sense of the crab's pincers only has significance in the sense of the life functions of crabs.

87 W. Strombach (1970) gives a review.

88 So, for example, R. Carnap (1966).

89 Thus, for example, the marsupial is defined as a "mammal with embryonally attached coracoid, with a pair of ossa marsupialia ..." and so on. The decision as to whether the range and contents of terms are correlated positively or negatively, depends on whether the term is understood extensionally (as the sum of its characters) or intensionally (according to its import).

90 This is the case for extensional determination.

91 In addition, note 57 above.

92 For example, consider the following series of concepts: atlas - cervical segment of the vertebral column - supporting apparatus - organism - species - animal kingdom - biosphere - earth - solar system - cosmos - matter - substance. Or: atlas - articular surface - triangle - straight line - point.

93 For example, in a structure with 10 sub-characteristics in 10^4 species times 10^7 individuals times 10^9 generations. The chance probability for a repertoire of two alternatives in each case amounts to $(1/1024)^{20}$, that is, about 10^{-60}.

94 Everything that can be inferred from such masses of the Cheops pyramids has been portrayed by Däniken, for example, with unintentional comicality.

95 The most important advocates of the synthetic theory are Th. Dobzhansky (1951), J. Huxley (1942), E. Mayr (1967), B. Rensch (1954) and G. Simpson (1964).

96 See R. Riedl (1976).

97 In E. von Holst (1969), who stresses that it deals primarily with the question of processing mechanisms which are intended to even out the distortions in perspective caused by the anatomy of the eye.

98 Further examples in E. von Holst (1969).
99 He who guesses aimlessly must pose, on average, about a thousand questions, but he who proceeds hierarchically can get by with, at the most, 11 questions ($2^{11} = 2048$).
100 This has been clearly presented, especially by E. Lenneberg (1972).
101 In which case, especially in the social sciences, there is a controversy about the necessity for hierarchy in societies ("classless" society, and so on). This theme is dealt with, for example, in R. Riedl (1976, Chapter 9).
102 See Kant: "Critique of Pure Reason".
103 This was first established in detail by K. Lorenz (1941), summarised in K. Lorenz (1973).
104 The concept of analogy is explained in the glossary.
105 The body symmetries are a product of evolution, whence from radial symmetry (e.g. Hydrazoa) via disymmetry (e.g. Ctenophora) to bi-lateral symmetry (e.g. Vertebrata) the symmetrical units decrease from many, through four, to two. Correspondingly the space axes increase from one, then two, to three. And corresponding to this is the development of the semi-circular canals in vertebrate animals (cf. Fig. 36).
106 For with the speeds and distances that we can achieve, insight into a multidimensional space, for example, is not necessary.
107 Empiricism and nominalism are explained in the glossary. Phenetism is derived from 'numerical taxonomy' and is an application of nominalism to biological structure research.
108 For a brief explanation of reductionism and behaviourism, see glossary. For social Darwinism, see H. Koch (1973).
109 See glossary.
110 In K. Popper (1972).
111 So perhaps by the switching on of innate killing inhibitions when using firearms. This problem has been dealt with by K. Lorenz (1963).

Chapter 4

1 D. Hume (1748) section VII part II, quoted from J. Wickert (1972; p.119ff.).
2 Cf. W. Staudacher (1942).
3 A general outline on this is given in the "Fischer-Lexikon", volume 11 (philosophy); see A. Diemer and I. Frenzel, ed. (1967).
4 This quotation and the historical background in E. Oeser (1971, p.93).
5 By animism (from the Latin, anima = soul) is meant the idea of the animation of all phenomena, whereby these "souls" are also causally responsible for all events.
6 See glossary.
7 David Hume, in contrast to I. Kant, understood the principle of causality not as cognitive a priori but described our idea of the necessary linkage of two or more events as the result of an (individual) accustoming process.

From the perception of the succession (post hoc) one could not logically draw conclusions about a causal connection (propter hoc).

8 He wrote, for example, in the "Prolegomena": "I freely admit: the memory of David Hume was that which for me, many years ago, interrupted my dogmatic slumbers" (A 7ff.), and in the "Critique of Pure Reason" he described Hume as one of the great "geographers of human reason".

9 Aristotle's causal concept figures in his "Metaphysics".

10 This situation is described in detail in R. Riedl (1978/79). "Intention" in the case of the dwelling of the beaver or the quiver fly larvae is genetically laid down, as with all living structures, and corresponds to the requirements of the formal and final causes as will be indicated in detail in the places named.

11 Those philosophers are intended who see their studies chiefly as interpretations of Aristotelian writings.

12 See glossary.

13 The difference between natural sciences and the humanities is often attributed to their differences in methods in which case the so-called "historical method" refers to the commentator's interpretation. Cf. E. Wentscher (1921).

14 Theophrastus, a pupil of Aristotle, had emphasised the primacy of the efficient cause, but only modern science has systematically readjusted itself, in that the question as to "why" was replaced by the question of "how". Cf. H. Sachsse (1967).

15 This separation into Arts and Sciences was specially effected by W. Dilthey (1933), whilst W. Windelband (1894) sought to make the term of natural science more precise.

16 Compare the developmental physiology of A. Kühn (1965) or F. Baltzer (1955). Baltzer established that, in the processes of ontogeny, "another and quite unknown form of determination was operating, a special nexus organicus" (in the sense of N. Hartmann).

17 For vitalism, cf. H. Driesch (1909) and the explanation in the glossary.

18 The following examples may illustrate this. In mathematically ideal billiards with eight balls, each a metre distant from each other, it is in principle impossible to predict whether the seventh ball will still hit the eighth (if the first is pocketed). The quantum theoretical uncertainty of the surface molecules exceeds by the eighth power the diameter of a billiard ball. Or: if a mutation is caused by a quantum transition, then this chance event acts as far as the phenotypical change of a characteristic.

19 One thinks of the numerous, at times contradictory, positions in vitalism and mechanism, in which sometimes only one form of cause is assumed to exist. Even N. Hartmann (1964) considered causality and finality as opposites. For the causal problem, compare also R. Kaspar (in press, 1980).

20 For example, one can describe an anti-particle mathematically as if it moved from the future into the past. For introduction of a time direction

becomes relevant only if it involves such large quantities of particles that order and disorder processes participate. Cf. E. Lüscher (1978).

21 The entropy concept is explained in the glossary.

22 N. Tinbergen (1951) has described these connections. A survey is presented by I. Eibl-Eibesfeldt (1978).

23 See K. Lorenz (1973, p.196ff.).

24 K. Lorenz, in particular, has repeatedly described this method of space presentation for abstract thought, most recently in (1973, Chapter 7). Cf. the contributions in H.-W. Klement (1975).

25 The expression "ratiomorphic" was first used by E. Brunswik (1955, 1957); see glossary.

26 The details of this were discussed by B. Rensch (1965).

27 In K. Lorenz (1973) and K. Popper (1974).

28 In 1904 this horse of Herr W. von Osten created a great sensation throughout the whole world. Similar "abilities" were possessed by K. Krall's horses, which could read various letters and could extract cube roots. Amusing descriptions are given by B. Grzimek in H. Friedrich (1968; pp.53-63) and P. Watzlawick (1976).

29 Very informative investigations on the linkage between cause and effect have also been made in psychology (A. Michotte, 1966).

30 In E. Mach (1905). Presented in context by E. Oeser (1976, vol.3; p.110).

31 K. Popper also speaks of this marvel in the foreword to "Objective Knowledge" (1972).

32 K. Lorenz, in P. Weiss (1971; p.231).

33 These are detailed in R. Riedl (1975, Chapter VII).

34 On this, among other things, depends the psychological effect of the joke, as S. Freud (1958) has shown.

35 The information content of the telegram may be given in "bits" (digital yes/no decisions), in which the number of bits corresponds to the required effort, for coding each sign used from a given repertoire.

36 If the repertoire consists of 32 signs (letters, punctuation marks and spaces), then every sign sent contains the information of 5 bits ($2^5 = 32$). The probability of getting it right by chance amounts therefore to $2^{-5} = 1/32$. This number is raised to the power of the number of signs sent.

37 For in the 10^{17} seconds that the cosmos has existed, 10^{44} attempts per second would have been necessary to allow this telegram to have arisen by chance.

38 In K. Lorenz (1974a). For example, in confusing the causal directions, one can mistake a current-driven compressor for a current aggregate driven by a piston motor or a water mill for an anchored river craft with stern paddle wheels.

39 See glossary ("analogy").

40 The phenomenon of mimicry (see glossary) has been fully described most recently by W. Wickler (1968).

41 The homology concept (see glossary) was introduced into biology by

Oken, the type concept by Goethe.

42 In R. Riedl (1975).

43 Morphological type is meant (cf. Chapter 3).

44 O. Koenig (1970) has shown this for cultural ethology using the example of development of uniforms; I. Eibl-Eibesfeldt (1978) has clarified the relationships.

45 For details see R. Riedl (1978/79).

46 The epistemological relations are described by E. Oeser (1978).

47 M. Planck and E. Mach have stressed this. Compare E. Oeser (1976, vol.3; p.121).

48 With increasing experience, the probability level of our foresights increase to astronomical proportions; for homology of the vertebral column amounting perhaps to the chance probability of $10^{-30,000}$.

49 For the formal simplicity of the propositions also increases as in the equation $E = mc^2$.

50 A. Einstein's views against the probability concept in quantum theory are given in the correspondence with M. Born (A. Einstein and M. Born, 1969).

51 This is shown in R. Riedl (1976) for all complexity layers.

52 In M. Eigen and R. Winkler (1973/74, p.113).

53 For example, the conservation principles of mass, energy, of impulse or of the centre of mass.

54 B. Hassenstein (1965) gives a series of examples for this. The "artificial turtle" would be an application of this rule technique (H. Zemanek, 1968).

55 See Kant: "Critique of Pure Reason", B 106 (the table of categories).

56 See glossary.

57 In E. Oeser (1976, vol.3, p.27ff.).

58 This becomes clear if, for example, one tries to describe a complicated machine, where each element can be understood in the whole context, but this can become obvious only after the description of all the elements.

59 Numerous examples in S. Freud (1958).

60 Quoted from P. Weiss (1971; p.82). J. Forrester developed the first computer model for representing the world-wide limits of growth.

61 Cycles of this kind always occur wherever the effects of a cause on the whole system react again on this cause.

62 In J. Galbraith (1970; p.11).

63 Thus it emerges, for example, that the decline of the reindeer in Lapland can be attributed to increasing the height of chimneys in England or that the decimation of Antarctic penguins was partly caused by excessive doses of DDT in North America.

64 Compare with this D. Dörner (1975); see also D. Dörner and F. Reither (1978).

65 One should refer to F. von Hayek (1952 and 1979).

66 The problem for our times has been excellently outlined by K. Lorenz (1974c). In this connection the papers by A. Huxley (1966) or G. Orwell

(1949) can be greatly recommended. Among others, B. de Jouvenel (1970) and E. Schumacher (1973) have striven for a solution of this world-wide problem.

67 Quoted from H. Schwabl (1958).

68 The natural history of consciousness can be studied in K. Lorenz (1973), H. von Ditfurth (1976) or R. Riedl (1976).

69 The so-called pre-Socratic philosophers were concerned with the problem of the "real" cause of all phenomena; Thales found it in water, Anaximenes in air, Heraclitus in fire, Anaximander in the "unlimited" (ἄπειρον), Parmenides in "being". Aristotle spoke of several causes.

70 In K. Lorenz (1973; p.29).

Chapter 5

1 Quoted from B. Russell (1946; p.637) and K. Lorenz (verbal communication).

2 In ordinary as well as in scientific language, the term "Seele" is often restricted to the human soul (mind). Rightly so if by that is meant the reflecting self-consciousness, which in its developmental history might appear in man.

3 The big bang in astrophysics means the events with which the existence of the Universe began (about 17 thousand million years ago). For further information, see H. Störig (1972) or S. Weinberg (1977).

4 It is interesting that the understanding of a cause depends on the evaluation of the circumstances, and this is transferred (in the sense of a moral judgment) to objects as if these had a free will.

5 Even this inaccessibility of purpose led Aristotle to interpret the purpose-directed event by an inner "entelechy", which, like the purpose itself, seems difficult to comprehend. Subsequently the entelechy concept was taken up especially by vitalism (see glossary), after it had previously played a central role in Mediaeval Christian philosophy. Here entelechy was equated with the original cause, namely God.

6 John Duns Scotus (ca. 1270-1308) was a Franciscan friar on the boundaries between high and late scholasticism.

7 In the paper "On the use of teleological principles in philosophy" (1788) I. Kant wrote, ". . . that in Nature there must be purposes that no man can comprehend a priori."

8 As set out in Chapter 4, with the beginning of modern science, purpose was renounced as a means of explanation because it seemed that efficient causes sufficed for physical phenomena (such as terrestrial and celestial mechanics).

9 See G. Hegel (1806).

10 See glossary.

11 H. Bergson has described this in detail in his main work "Creative Evolution" (1912).

12 A comparative survey on this subject in C. Kernig (1968).

13 Quoted according to Borzenko (1963) from C. Kernig (1968, vol.2, p.510). See also H. Hörz and C. Nowinski (1979).

14 For example, in L. von Bertalanffy (1968), W. Heisenberg (1966 to 1976), K. Lorenz (1973), M. Planck (1965), P. Weiss (1971) and D. Campbell (1974a).

15 See R. Riedl (1976).

16 This subject is dealt with in R. Riedl (1976, Chapter 4); from the specialist literature referred to, M. Calvin (1969), C. Ponnamperuma (1972) and H. Urey (1952) are recommended.

17 The theory of the so-called "hypercycle" (see glossary) is described by M. Eigen and R. Winkler (1975) and P. Schuster (1972).

18 This interaction of purpose and organisation has its counterpart in genetic patterns, in the epigenetic system and in the functional relation of induction; (all the terms are explained in the glossary).

19 Compare C. Kernig (1968).

20 In K. Lorenz (1973).

21 As K. Lorenz has described, conscious, reflective thinking developed, starting from simple reactions like taxis and instinct (see glossary), through the "teleonomic modification of behaviour" and training by rewards, and ultimately by way of concept formation, insight and curiosity behaviour to imitation and tradition, whence consciousness became a complex, even superindividual phenomenon. A summarising presentation in R. Riedl (1976, Chapter 8); collected contributions in H. -W. Klement (1975).

22 In B. Rensch (1973).

23 Quoted from B. Rensch (1973; p.202).

24 This and numerous similar examples in J. von Lawick-Goodall (1971).

25 Muscle tissue is composed of fibrils which consist of actin and myosin molecules. The latter form a myosin protofibril for every 400 molecules.

26 See Chapter 1 (note no.73).

27 Even this is only of limited validity, namely, in an average biologically relevant area.

28 So every great new artistic creation, in spite of its uniqueness, is seen as a "child of its time". What would Aristotle be without Plato, or Leonardo da Vinci without the Renaissance? We all stand on the shoulders of ancestors.

29 Such striking circumstances in the prehistory of human culture are described, for example, in K. Narr (1961).

30 The popular account by T. Prideaux (1973) provides a survey.

31 The ability to get over the prejudice of this "magic thinking" was examined by psychology with various problem-solving tasks. One example: the problem is to seize with one's hands two ropes that are hanging down but are not simultaneously within one's grasp. The solution is to cause one rope to swing about a pivot, to seize the other and hold it fast until one can catch the first rope as it swings back; cf. F. Klix (1976; p.656).

32 Quoted from I. Kant (1804).

33 Thus, I. Kant (in the "Critique of Pure Reason"): "The most essential and the most important . . . however, is that the notion of final cause in nature, which separates the teleological judgment of it from that according to general, mechanical laws, belongs to the power of judgment but not to the understanding or to reason."

34 Compare "Critique of Judgement" (§ 72).

35 With N. Hartmann (1951) or J. Monod (1971), final and causal considerations are difficult to combine. Of course, C. F. von Weizsäcker, K. Lorenz, L. von Bertalanffy and others represent contrary interpretations.

36 Quoted from the "Critique of Judgement" (§ 68).

37 Quoted from R. Eisler (1930; p.623).

38 Quoted from R. Eisler (1930; p.626).

39 Quoted from the "Critique of Judgement" (§ 72-74).

40 Quoted from the "Critique of Judgement" (§76).

41 Quoted from the "Critique of Judgement" (§71).

42 Quoted from I. Kant (1788).

43 This subject is dealt with in detail in R. Riedl (1978/79).

44 Compare Chapter 1, note 23.

45 P. Teilhard de Chardin (1959) has expounded this position from his viewpoint.

46 As already mentioned in Chapter 1 (see note 18), it was always one of the central philosophical problems to bring into harmony the determinateness of the world with the freedom of man.

47 This explanation was the reason why Einstein withdrew from the Copenhagen interpretation of the quantum theory. The quotation from M. Eigen and R. Winkler (1973/74; p.113).

48 In R. Riedl (1976, especially Chapter 3).

49 In W. Stegmüller (1969; p.518).

50 Quoted from W. Stegmüller (1969; p.519).

51 See R. Riedl (1975).

52 For example, if the development of two articular surfaces belonging together is coupled genetically. If every mutant had to "wait" until the other bone changed correspondingly, then a million times the effort would be required for evolution. Cf. R. Riedl (1975, 1976, 1977) or R. Kaspar (1978).

53 The Bilateria are animals with a symmetry plane, that is all with the exception of Protozoa, Coelenterates and sponges. The starfish, for example, shows a secondary transformation. Body cavity animals are those which show a secondary body cavity, a coelom. The placenta animals are a sub-class of the mammals. Primates (master animals) are a class of the placenta animals, to which man also belongs.

54 E. Mayr (1970) gives a more detailed explanation.

55 The phenomenon of orthogenesis is described more fully in R. Riedl (1975, particularly pp.318ff.).

56 Thus, for example, an extreme parasite from the group of the Cirripedia, namely the Sacculina, has not left its sub-class. Even the primitive bird Archaeopteryx with all its derivatives, the birds, has not left the super-class of the Sauropsidae and the sub-branch of the vertebrates.
57 See "Critique of Judgement" (§ 75).
58 Quoted from I. Kant (1804).
59 See R. Riedl (1976).
60 Compare note 13 above.
61 In Kant: "Critique of Judgement" (§ 81).
62 See N. Hartmann (1964; p.507).
63 For example, in Kant: "Critique of Judgement" (§72-74).
64 K. Lorenz (verbal communication).
65 Quoted from R. Eisler (1930; p.628).
66 Quoted from R. Eisler (1930; p.628).
67 On this subject the writings of G. Orwell (1949) are recommended.
68 Those who wish to become informed on this from an outstanding, close-to-life account, should refer to "Cultural History of Mankind" by W. and A. Durant (1960).
69 With Aristotle, the existing works written after those on "nature", were denoted as the "meta-physical", those "after the natural".
70 Neoplatonism, founded by Plotinus (205-270), was in its prime up to the Middle Ages, but the tradition continued as far as the Anthroposophy of R. Steiner.
71 See K. Marx and F. Engels (1846).
72 For this mechanism of the social construction of realities, see P. Berger and Th. Luckmann (1969) or P. Watzlawick (1976).

Chapter 6

1 Quoted from G. Vollmer (1975; p.172) and E. Oeser (1976; vol.3, p.119).
2 Demands from G. Vollmer (1975; pp.185-6). Compare also the detailed treatment of the various centres of gravity in evolutionary epistemology. F. Wuketits (1978) gives a survey.
3 For detail, see K. Popper (1959, 1972).
4 In H. Albert (1968, p.13), quoted from G. Vollmer (1975; p.25).
5 The solution of Descartes will be remembered: "Cogito, ergo sum" (1641); compare also B. Pascal (1645), A. Einstein (1949). In that connection there is also the example of K. Lorenz (1959), that the mechanism of counting functions like the shovel of a dredger, and basically is correct only if the simple idling is counted. As soon as real objects are counted together, the statement $1 + 1 = 2$ is incorrect because the objects are never identical.
6 For, as we have mentioned, the logical truth of a statement is no guarantee at all of its factual truth.
7 This is the age of life on earth as well as the age of the cosmos.

8 See M. Eigen and R. Winkler (1975).

9 Thus there is an algorithm that provides repetitive calculating steps for each division, with which this can be carried out as accurately as desired. The same holds for the process of taking roots, and so on.

10 This has been realised for the biological domain in R. Riedl (1975), and for evolutionary phenomena in general in R. Riedl (1976).

11 The first quotations are from E. Oeser (1976, vol.3; p.119ff). There, the detailed description and epistemological basis can be consulted. The last quotation is from E. Oeser (1979, p.24). Compare also W. Whewell (1840, part I; p.26). W. Ostwald (1898; p.31), P. Volkmann (1913; p.26) and E. Mach (1921; p.260).

12 Such a non-redundant cosmos would contain each form of a certain regularity only once, which would not enable us to distinguish it from chance. However, order is always composed of the content of law and its application (redundancy) (see R. Riedl, 1975).

13 Again, cf. R. Riedl (1975, 1976) as well as H. Sachsse (1967).

14 In D. Campbell (1974; p.418).

15 In S. Pepper (1958; p.106).

16 In G. Simpson (1963; p.84).

17 In H. Mohr (1967; p.21).

18 These last quotations are from B. Rensch (1968; p.232), H. Sachsse (1967; p.32), A. Einstein (1972; p.119ff.), and S. Pepper (1958; p.106).

19 This dispute is about the question whether there is a common principle underlying the material and the spiritual (monism), or whether it involves two fundamentally different manifestations of reality, which are not interconvertible (dualism). On this, see B. Rensch (1968) and R. Kaspar (1980a). We do not accept the dualism still represented by J. Eccles.

20 Quoted from W. Stegmüller (1971; p.18).

21 From D. Campbell (1974; p.422).

22 According to W. Stegmüller (1974; pp. 1 and 2).

23 In D. Hume (1748). Quoted after G. Vollmer (1975; p.6ff.).

24 From A. Einstein (1972; pp.115 and 119). Einstein includes the animals.

25 Refer to the glossary for an explanation of these terms.

26 Quoted from W. Stegmüller (1954; p.535).

27 The expression “category” is derived from the law and means the accusation “on the place of judgment”(κατ᾽ αγοράν).It was then taken over by Aristotle into his philosophy, in the sense of a principle of thought, as later treated in detail by I. Kant in the so-called critical writings.

28 This quotation is from K. Lorenz (1941; p.99), a work in which, for the first time, an evolutionary interpretation of Kant’s a priori is given. Moreover, at this time K. Lorenz was ordinary professor in Königsberg in the chair formerly occupied by I. Kant.

29 Quoted from I. Kant (1770; 5 and 1790a; 1, Abs.C). In between there appeared the great critical writings (1781, 1790).

30 The account by D. Campbell (1974; p.441ff.), in this respect, is also a mine

of valuable information.

31 There is an explanation of these ideas in the glossary. Also compare the discussions on this theme in Chapter 1.

32 See J. Eccles (1973) and the further references there.

33 See J. W. von Goethe (1790) and A. Remane (1971). For criticism on morphology, see B. Hassenstein (1951, 1958), for discussion on the concept of idea in morphology refer to R. Kaspar (1977).

34 The fundamental problem for morphology lies in the circumstance that it was not in a position to describe the methods of its procedures. A first attempt at solution is found in R. Riedl (1975); likewise for the comparison theorem in R. Riedl (1976). A comprehensive account of the subject of concept formation is in preparation (R. Riedl and R. Kaspar: "Biology of concept formation").

35 Quoted from K. Lorenz (1943; p.322) and G. Vollmer (1975; p.105).

36 For the cause of homology and type see R. Riedl (1975), for form perception K. Lorenz (1959).

37 Anaxagoras, an ancient Greek natural philosopher, lived from 488-428 B.C.

38 Compare N. Hartmann (1964; e.g. p.507ff.).

39 For this, see the quotation in note 33, Chapter 5.

40 This subject is dealt with in detail in R. Riedl (1978/79), and in Chapters 4 and 5 of the present book.

41 Dialectically here means this mutual relation of causes and effects; to be distinguished from its meaning in dialectical materialism, which is not "dialectical" insofar as it only considers one causal component, namely that from the simple to the more complex.

42 This sorting into the four decades refers to the first editions of the works in question. As far as possible, however, in the references we generally prefer to indicate more recent editions, so as to assist the reader to further information. The works concerned here are: K. Lorenz (1941); L. von Bertalanffy (1955); D. Campbell (1959); N. Chomsky (1968); H. Mohr (1967); J. Piaget (1974); B. Rensch (1968); E. Lenneberg (1972); J. Monod (1971); K. Popper (1972); E. Oeser (1976); A. Shimony (1971) and G. Vollmer (1975).

43 See S. Freud (1940), C. Jung (1954); the quotation is from E. Neumann (1974; p.7). compare especially the formulation in K. Lorenz (1973).

44 In one of the joint seminars of the University of Vienna (1977) on evolutionary epistemology.

45 It will be remembered that Goethe allowed his Faust to despair of this viewpoint; cf. Faust I, line 365.

46 Compare with this note 130 in Chapter 2.

47 The corresponding places may be found in D. Hume (1748), I. Kant (1781; B.180), Hubert Rohracher (1953; p.8), N. Chomsky (1971).

48 Quoted from the introduction to J. Locke (1690).

49 From B. Russell (1963).

50 See M. Gazzaniga (1970), R. Sperry (1970a, 1970b), as well as the more recent survey in K. Walsh (1978).

51 Especially E. Oeser (1976, vol.3).

52 A survey is given by J. Eccles (1973).

53 For the many informative examples (in hidden places), refer to the thesis by Harald Rohracher (1948).

54 Compare I. Pavlov, the summary by W. Deglin (1976) and the essay, informative in many respects, by K. Lorenz (1959).

55 The ratio of hours of instruction in mathematics plus Latin to drawing plus music in an Austrian Grammar School is as 2:1. Even more marked is the difference if one compares the number of teaching hours with that of the hours of homework expected by the school.

56 The corresponding numbers are, for example, available in the annual reports of the German Research Council (Jahresberichte der Deutschen Forschungsgemeinschaft) which appears annually as "Activity Reports" (vol.I) and as "Programmes and Projects" (vol.2).

57 It is gratifying, however, to find that a few institutions promoting education now recognise this decline. For example, the "Studienstiftung des Deutschen Volkes" (Bad Godesberg) has used the paper by H.-R. Duncker (1978) as an occasion of multidisciplinary discussions.

58 Compare D. Dörner and F. Reither (1978), as well as Chapter 4, p.112, in the present volume and Fig. 48.

59 Social psychological information may be found in P. Berger and Th. Luckmann (1969) and in P. Watzlawick (1976), that for the immunisation of scientific theories in H. Albert (1968) and K. Popper (1972); a general survey is given by R. Riedl (1976).

60 As will be remembered, ontological reductionism maintains that every more complex phenomenon is nothing other than the interaction of its simplest sub-systems (see glossary). The so-called "tabula rasa" position emerges from the view that at birth the brain is a "clean slate", so that any sort of thought content is a product of personal experience (see glossary, "empiricism").

61 Thus, positivism (see glossary) reduces epistemology, indeed the whole of philosophy, to formal logic and logistic, social Darwinism reduces biology to a misunderstood theory of selection, behaviourism reduces (see glossary) psychology to the statistics of reactions, phenetics reduces morphology to the counting of characteristics and dogmatic genetics permits no reactions of phenomes on genes, which may be correct chemically but is otherwise false; for how can genes learn anything without retroactions?

62 Some say that this view has been deliberately prevented although the experts have clearly defined it for us. Compare J. Forrester (1971), J. Galbraith (1970), B. de Jouvenel (1970) and E. Schumacher (1973).

63 See F. Seitelberger (1975; p.9).

64 This parallel with parasitism in the animal kingdom is enlarged upon in R. Riedl (1976).

65 This has been formulated for technology by I. Rechenberg (1973).

66 For a survey of movements contemporary with G. Galileo and E. Haeckel see J. Hemleben (1964 and 1969). For the comparative problems of the three Copernican revolutions, see R. Riedl (1979).

67 From the well-known saying of A. Schweitzer: "Ich bin Leben, das leben will, neben anderem Leben, das auch leben will." ("I am life that wishes to live alongside other life that also wishes to live").

68 We refer, for example, to A. Runge. Compare J. Huxley (1954).

GLOSSARY

An arrow → signifies reference to other terms mentioned in the glossary.

Algorithm. An algorithm is a calculation process determined by rules, and it makes possible the solution of a certain class of problems through cyclically repeating operations (e.g. Gauss's algorithm for the solution of systems of linear equations). In this way, frequent repetitions of a small number of operations can lead to the optimisation of a solution, as, for example, in division. Moreover, an algorithm in logic indicates a special method for determining the value of calculable functions. Here, therefore, it is regarded as a method of decision. Everywhere in the present volume, the former, more general meaning of algorithm is intended.

Analogical inference. This arises when from experience of particular objects with known properties, we infer to other similar objects with partially unknown properties. For example, when, on the basis of one's own experience, a subjective feeling is attributed to higher animals too. This inference is highly unjustified, as is the assumption of subjective experience with fellow creatures, i.e. cannot be proved. However, like → induction, it extends not experience but expectation. It arises if, for example, we expect to find the same contents between similar book covers.

Analogy. In biological structural research, analogy is the term used for that form of similarity which arises by reason of independent adaptation to the same environmental conditions. Therefore the cause of analogous similarity does not lie in the system conditions of the structures themselves but outside them. In biology, the wings of an insect, a flying saurian or a bird, for example, are analogous. Moreover, there is a distinction between functional and chance analogy (→ similarity field). See also Figs 41, 42.

A posteriori. Translated, the expression means "from later". Knowledge a posteriori signifies that which is achieved in retrospect simply on the basis of experience. The distinction is thereby emphasised from knowledge which must be present for the experience to be possible. → a priori.

Appetitive behaviour. → instinct.

A priori. This expression literally means "from earlier". In epistemology, it signifies the knowledge that must be given to the subject desiring to know, before each experience. For human beings, therefore, clear a priori views are perhaps the three dimensions of space, the categories of causality, time, and so on. Biological epistemology is able to interpret the a priori (for the individual) as a posteriori for its phylum, for the assumption of, say, only three space dimensions is no doubt the result of a phylogenetic gain in experience and, in that sense, is not "absolutely evident".

Association. → conditioned reflex.

Axiom. Every branch of knowledge depends on certain assumptions or prerequisites, whose validity cannot be explained in themselves. If, for example, Euclidean geometry begins with the shortest link between two points being a straight line, then this basic proposition cannot be derived from geometry. These ultimate presuppositions of a science, that cannot be established independently, are called axioms. The entire administration of justice, for example, depends on the axiom that there is free will and personal responsibility.

Behaviourism. A school of psychology found especially in America, the programme of which involves research into ways of behaviour and is limited to counting and measuring the "observable". In its theoretical concepts it depends essentially on reflex theory and it denies the existence of inborn ways of behaviour in the sense of → instincts. The error of this position lies in the obviously erroneous view that the reflex should be the only element of animal and human behaviour.

Category. This expression goes back to Aristotle. In Kant's view, categories are "logical conditions of experience", hence those principles of thought which facilitate an ordered point of view in consciousness. The number of categories has been variously given, depending on the author. Examples of categories are: quantity, quality, relation, time, space, causality. Categories are not derivable from reason itself and therefore are given → a priori.

Chance. → determinism, → indeterminism.

Conditioned reflex. Every reflex of an organism has an unconditioned release, corresponding to the natural environment; a change in illumination alters the size of the pupils, and the like. If, shortly before the natural stimulus, (A), a preferential stimulus, (B), is repeatedly applied at a given time, then an association is formed between the two stimuli (B-A), so that eventually the first stimulus (the conditioned release) suffices to permit the reflex to take place. This is then denoted as a conditioned reflex: B-A reflex leads to B-reflex. See also Figs 22 and 23.

Constancy performance. To facilitate classified perception contents and orientation in time and space, particularly in higher organisms, highly complicated processing mechanisms are necessary in the central nervous system and in the sense organs. The constancy performances due to them permit, for example, the recognition of an object moving before the eyes as constantly the same although it produces a large number of different retinal images because of its movement. Other constancy performances permit us, for example, to apprehend a white object always as the same colour despite the most varied kinds of illumination. See also Figs 24 and 25.

Deduction. An inference pattern in formal logic which relates the truth content of general statements to other special propositions by means of definite rules. An example of the simplest kind of deductive conclusion would be:

(a) all logicians infer deductively
(b) N.N. is a logician
(c) N.N. infers deductively

(a) and (b) represent the premisses, (c) is the conclusion, which here (as against → induction) does not overstep the bounds of the premisses.

Determinism. This denotes a basic philosophical view that postulates an unequivocal cause for every event. In determinism, chance is understood subjectively, namely, as caused by inadequate insight into the causal relationships. In physics, e.g. Albert Einstein represents determinism; in biology, Bernhard Rensch, among others. See also → indeterminism.

Empiricism. Theoretically, empiricism lays the basis of all knowledge in the awareness and experience of the subject. Out of this, there often arises the so-called "tabula-rasa" notion, which ascribes every capacity for knowledge and thought to individual experience. The senses as the only source of knowledge, are formulated in the Aristotelian proposition "Nihil est in intellectu quod non prius fuerit in sensu" (that is, "nothing is in the understanding that was not previously in the senses"). See also → rationalism.

Entropy. The term "entropy", introduced by R. Clausius in 1865, denotes a quantity of state of a thermodynamic system. It may be calculated by taking the system reversibly from an arbitrary initial state into another state, by determination of the heat added (δ Qrev) divided by the absolute temperature (T) and summation of the quotients. Expressed in the formula:

$$S - S_o = \int \delta\, Q_{rev}/T \text{ or } dS = \delta Q_{rev}/T$$

For irreversible processes, $dS > \delta\, Q_{irr}/T$

Entropy may be illustrated, as E. Schrödinger put it, by the disorder slowly arising on a desk, if it is not continually cleared.

Epigenetic system. In biology, the term epigenetic system denotes the totality of all regulative interactions in the genome of a biological unit (e.g. species, family, and so on). Thus it deals with a dynamic ordering principle, that among other things, ensures that phylogenetically older conditions are recapitulated during embryonal development. Since, in the course of evolution, every experience of the genome is taken up into it, its own history is reproduced. The possibility of atavistic repercussions is a proof for it.

Evolutionary epistemology. In contrast to the various philosophical epistemologies, evolutionary epistemology attempts to investigate the mechanism of cognition from the point of view of its phylogeny. It is mainly distinguished from the traditional positions in that it adopts a point of view outside the subject and examines different cognitive mechanisms comparatively. It is thus able to present objectively a series of problems that are not soluble on the level of reason alone, indeed what we are attempting in this book.

Fulguration. This term was introduced into biology by Konrad Lorenz. By it we understand the phenomenon of the fusion of two or more systems into a new unit which then shows qualitatively different properties from its elements. For example, echo sounding arises from the combination of ultrasonic sounding and the ability to calculate these frequencies acoustically. In the evolution of living things, new system regularities continually arise; so the quality that is specifically human is formed by a synthesis between the idea of space, the gripping hand, erectness, curiosity behaviour and the development of speech. Life itself is a specific system regularity which is contained in none of its physico-chemical properties alone.

Heuristics. In contrast to (deductive) logic, heuristics seeks to develop a method which facilitates the search for useful hypotheses. Its approach, therefore, is essentially inductive. Heuristics, basically, has two tasks: (1) the problem of hypothesis formation by means of definite "search rules", and (2) the problem of hypothesis evaluation, which can be treated by formalisable methods. With hypothesis formation, chance participates insofar as the expected regularity cannot be based solely on previous experience.

Homology. That form of biological similarity which is established by the same regularity of → epigenetic systems. It is distinguished from → analogy in that the cause of the similarity itself lies within the systems. For example, the skeleton structures of the anterior extremities of the whale, bird, bat, horse and man are homologous. Homologous similarity remains in spite of differing environmental requirements. The homology is recognised from divergent, harmonious → similarity fields. See also Fig. 26.

Hypercycle. In the pre-cellular phase of evolution, short chains of nucleic acids and various proteins existed at first independently of each other. Each nucleic acid represented a small, self-reproducing cycle (positive ↔ negative), until several such cycles, functionally coupled by proteins, fused together into a hypercycle. From this cooperative interaction between "legislative" (nucleic acid) and "executive" (protein), according to Manfred Eigen's theory, there should result the genetic code, in which it would become possible to form sufficiently long nucleic acid chains and, at the same time, to lower the frequency of errors.

Hypothetical realism. → Evolutionary epistemology depends, among other things, on the knowledge that our cognitive apparatus itself is a thing of actual reality to which it would adapt in the course of evolution. This leads to the decisive hypothesis that what our cognitive apparatus communicates to us about the world corresponds to something real. This position of hypothetical realism differs, however, from naive realism by the insight that objective knowledge is to be acquired only by the knowledge of regularities in the world picture itself. (Here we are to understand those physiological mechanisms which serve for acquiring knowledge.)

Idealism. The epistemological position of idealism assumes that the external world does not exist independently of the perceiving subject but only as an object of possible experience. In this respect one must distinguish the transcendental idealism of Immanuel Kant, which postulates (namely, in the thing itself) a reality that exists beyond and independent of experience. The logical consequence of idealistic epistemology would be solipsism, which ascribes reality only to the individual subject and interprets all other experience as the product of its power of imagination. In addition, causes are usually explained only as final.

Imprinting. A special case of the learning process, in which the learning content can be assimilated only during a short phase of development and from then on it is retained irreversibly. Some organisms learn the image of their parents or sex partners by imprinting. The physiological mechanism underlying this is open to every learning content, and hence suitable organisms can be imprinted very readily for experimental purposes on humans. Likewise man himself is imprinted, e.g. on the conditions of his civilisation. See also Fig. 9.

Inborn release mechanism (IRM). Every motor response of an organism to a stimulus of its surroundings is switched on by a physiological mechanism which exerts the function of a stimulus filter. It is comparable to a lock which opens only to a quite special key (stimulus). This stimulus-filtering mechanism is denoted as an IRM. For example, with the tick (Ixodes rhicinus) the detection of butyric acid and 37°C is sufficient to release the piercing reaction, which shows that the information that the IRM contains is extremely simplified and schematised.

Indeterminism. The basic philosophical attitude that emerges from the postulate that the course of events in nature, in principle, contains non-causal elements and that therefore "objective" or "real" chance exists. This view is frequently supported by interpretations of quantum physics which conclude that, because for certain events a cause cannot be established, such a cause neither exists nor can exist. See also → determinism.

Induction (in embryology). The transfer of information between two tissues, due to chemical substances during the ontogenetical formation of an organism, is termed induction. Thus, the eye-stalk transfers information on lens-formation to the skin of the head and the lens, information on the invagination of the eye-stalk. The inducing substances must be the same in large groups, as the transplantation experiments of Spemann first showed. The induction pattern for eye formation is valid, for example, for the whole sub-phylum of vertebrates.

Induction (in the cognitive process). In epistemology one always speaks of an inductive inference from the special case to the general, when a general assertion is concluded on the basis of certain individual experiences, the content of which oversteps the range of the premisses. Here, induction is not a logical but a heuristic process, since a logical conclusion, by definition, may not exceed the bounds of its premisses. It is induction, for example, if some specimens of an animal species are dissected and, from that, the anatomy of all individuals of that species is described. All natural laws are obtained inductively; they are tested by the aid of deduction. Inductive inference extends, not experience, but expectation.

Instinct. An instinct or a specific drive activity is a genetically fixed movement, the inducement of which is generated endogenously (i.e. without any release by external stimuli). The instinctive movement itself is initiated by an inborn release mechanism (IRM) and runs constantly in its species-specific manner. In the absence of any releaser, the threshold lowering proceeds until it becomes a so-called idling movement, the instinctive movement thus proceeding without any reference object. In many cases, instinct is associated with some preceding appetitive behaviour, a search for the biological releaser.

Kinesis. Movement; it has to do with simple ways of reaction which occur even with unicellular animals. It causes the organism to move faster as soon as it reaches unfavourable environmental conditions, but retards it in a favourable medium. The direction of movement is not thereby affected. Apart from kinesis in Protozoa and some Isopoda, there is an analogous mode of behaviour in mammals, perhaps with grazing ruminants. It is well known that a man seeking fungi behaves in a similar manner. See also Fig. 4.

Lamarckism. In the theory of evolution developed by J. B. de Lamarck (1809), it is assumed that individually acquired modifications react on the genetic material and so are transmitted directly to the next generation. This view was abandoned when it was recognised that a chemical reaction by the phenome with the gene was not possible. On the contrary, it should not be overlooked that a stochastic retro-effect by way of an alteration in the chances of mutants is actually possible. However, in that case it is not the medium but the organisation of the organism that reacts on the genetic material.

Materialism. The philosophical position that assumes that the basic reason for everything real is in matter is termed materialism. Its fundamentals were laid down in ancient hylozoism (animation of matter), but the further development was due to classical materialism (e.g. Lamettrie) leading to scientific materialism (e.g. Haeckel) and the so-called dialectical materialism (e.g. Marx). The characterisation of the numerous materialistic positions depends on the attendant concepts of matter. In general, even materialism, when seeking to answer the question of "original reality", must be regarded as metaphysics; "matter" is no less metaphysical than the "world spirit". In materialism causes are usually explained as merely efficient.

Metaphysics. Every philosophical position or attitude which employs assertions of which the truth content is not amenable to proof can be called metaphysics. In recent times, metaphysics became the science of the presuppositions of experience (Descartes), and with Kant, "the completion of all culture of human reason". Today, starting from positivism, numerous attempts are made to eliminate metaphysical questions from science, in which they are referred to as pseudo-problems (e.g. Carnap, Wittgenstein). On the other hand, there seems to be scarcely any field of human activity in which metaphysical aspects are absent. Every → inductive hypothesis or expectation goes beyond the pure realm of experience.

Mimicry. Mimicry is that extreme form of → analogy, in which the individuals of a species imitate exactly those of another species down to the fine details of their appearance. The imitating species thereby enjoys the advantage of being able to deceive its competitors in the actual features of the imitated species. For example, a sabre-toothed slime fish imitates a scavenger fish so deceptively that it is taken for a scavenger and pieces can be bitten out of the fins of other fish by it. See also Fig. 43.

Natural system. In biology, the natural system is understood to be that classification of organisms which represents their natural phylogenetic relationships. It is not designed exclusively for purposes of orientation (artificial system, e.g. Linné), but arises by the comparative appraisal of → homologous and → analogous features. The discovery of the natural system, therefore, resulted from the recognition of determinable similarities; and these similarities are explained by natural relationships. See also Fig. 44.

Nominalism. The old philosophical argument about whether reality is attributed to the general as it is to the special (to the individual concrete thing), reached its height in → scholasticism with the so-called controversy of universals. The position that asserts that the general has real existence only as regards the names given to things, is called nominalism. The question, among others, becomes of interest to biology where the problem is to decode whether the type of a species or a genus, and so on, really exists as well as the single individuals concerned.

Optimisation. → algorithm.

Order. In the physical sense, the concept of order can be derived from entropy. If this indicates the atomic disorder of a system, then the negative entropy (order) is given by its inversion: $N = k \cdot \log D^{-1}$ (k = Boltzmann's constant; D = atomic disorder). For biology, that description of order proves to be most convenient which considers it as the product from the lawlike content of a system times the number of its applications. The extent of order thereby increases both with regularity and with → redundancy.

Orthogenesis. This term denotes the rectilinear development running in trends of → trans-specific evolutionary processes. The cause of orthogenesis lies, on the one hand, in the internal system conditions of organisms themselves (→ epigenetic system) and on the other, in a relative constancy of the decisive environmental conditions. The best known example of an orthogenetic development is the evolution of the horse.

Phobia. A series of lower organisms respond to the onset of unfavourable conditions in the surroundings with a stereotyped avoiding reaction, which is termed a phobia. One of the best known examples is the phobic reaction of Paramecium. The amount of information contained in the phobia exceeds that of the → kinesis, inasmuch as the organism also learns something about the direction in which the unfavourable situation occurs. However, it contains no information as to the direction in which favourable conditions may be sought. See also Fig. 4.

Positivism. A philosophical attitude can be called positivistic if it is based on the assertion that the true source of man's knowledge is only that which is given (the "positive facts"). David Hume may be regarded as the founder of modern positivism and essential contributions were provided by the French encyclopaedists (e.g. d'Alembert). With increasing concentration on linguistics and logical problems, there arose neopositivism which received decisive suggestions from the Vienna Circle (e.g. Schlick, Carnap, Gödel). In addition, Ludwig Wittgenstein exercised much influence. However, no gain in knowledge is possible without experience → a priori and without expectations lying beyond experience; c.f. → metaphysics.

Pre-established harmony. The phenomenon of an ordered world constantly throws up the question of the cause of this order. In this respect, various philosophical positions, especially in connection with Leibniz, maintain that the harmony in nature may be derived from a pre-existent plan, in which the forms of later order were already pre-determined. The harmony between the elements of a totality, therefore, would not arise with the system but simply be the expression of pre-established harmony of a comprehensive world plan. The contrary point of view sees harmony developing with the systems themselves.

Ratiomorphic apparatus. The overall cognitive capacity of man is due to those physiological mechanisms of the central nervous sytem that is termed the cognitive apparatus. The accomplishments available to consciousness and self-reflection make up the system of reason. On the other hand, the ratiomorphic apparatus consists of those processing mechanisms which, as the phylogenetic precursor, represent the functional prerequisites of reason. This has to do with those unconsciously operating cognitive performances which are dealt with in this book from the point of view of four hypotheses.

Rationalism. In contrast to → empiricism, rationalism sees the true basis of all knowledge not in sense awareness but in the understanding. Since the evidence of the senses may be impaired by many kinds of deceptions, only reason with its → a priori forms of intuition and thought can ensure true knowledge. Rationalism had its peak in the seventeenth century with Descartes, Pascal, Spinoza and Leibniz.

Reafference principle. Orientated movement in space presupposes the ability to differentiate the sense perceptions which, for example, arise from its own movement from those that are caused by a change in the environment. The displacement of an object on the retina can be derived from an active movement of the eye or from the movement of the object itself. The reafference principle allows it to distinguish this; with the motor impulse that moves the eye, there is initiated simultaneously a corresponding message "with inverted sign" and is conducted into the processing centre, where it is compared with the starting information (the information from the retina). If the two messages are of equal size, they are blotted out and the object appears at rest in spite of the displacement on the retina. The colour constancy (→ constancy performance), inter alia, also depends on the reafference principle.

Reductionism. The explanatory tracing back of a phenomenon to the regularities of its component elements. For example, one tries to explain physiological phenomena by the laws of its underlying chemistry. Such a methodical reductionism is the fundamental approach of every natural science. Ontological reductionism, however, commits an error in asserting that a phenomenon is no more than the result of the interactions of its elements. It thus confuses method and reality. Water consists indubitably of hydrogen and oxygen, but its properties are qualitatively different from those of the two elements.

Redundancy. It is an essential prerequisite for the recognition of lawlikeness that this should occur repeatedly. The notion of redundancy indicates that portion of information which, in principle, could be omitted without affecting the information content. For example, in the present edition of N copies of this book, N-1 copies are redundant, since they do not extend its information content. Wherever a system that receives information has foresight, the information sent is redundant.

Scholasticism. The scholastic period in the history of philosophy began with the church fathers (e.g. Augustine) around 400 A.D. and lasted until the 14th century (e.g. Duns Scotus). Its philosophy, fashioned by Christian theology, was an ancilla theologiae, a "handmaid of theology"; it saw its main task in investing the contents of belief (say, proof of God, etc.) with reason. In the centre of the scholastic world view stands the universal stability of man's purpose and goals and the world directed towards God.

Similarity field. A domain of organisms or objects delimitable from its surroundings and sharing a certain set of features. The field may be harmonic (homologous features (→ homology) as in mammals), dispersed harmonic (functionally analogous features (→ analogy) as in animals' wings), or dispersed (accidentally analogous features (→ analogy) as in bell-shaped objects). See fig. 26, 41, 42 on p. 168, 175, 176.

Simultaneous coincidence. In the awareness or cognition of similarity fields, different constellations of features must be processed together. In principle, this involves an analysis of feature coincidences. Of these, simultaneous coincidence signifies that which is also denoted as abundance of characteristics; hence, in the objects of a similarity field, there are structures which are constantly perceivable at the same time. Thus, the second cervical vertebra in man is constantly recognisable from certain simltaneously occurring features. See also Fig. 27.

Solipsism. → idealism.

Sophists. This term was given to a group of pre-Socratic philosophers who first expounded philosophy to a broad public (for payment). Sophist (Greek: σοφιστής) means "master in knowledge" or "learned one". The most important Sophist was Protagoras ("Man is the measure of all things"), who lived from 480 to 411 B.C. From their fundamental criticism there finally grew a general scepticism which often led to empty debates and eloquent fallacies (sophisms).

Successive coincidence. As already mentioned in connection with simultaneous coincidence, the prerequisite of knowledge of similarity fields is also involved in successive coincidences. In this case it is a question of the features in the field which can be determined one after the other, like the second cervical vertebra in all mammals. Altogether, the processing and assessment of features rests on the product from simultaneous and successive coincidences, hence on abundance of characteristics times the number of individuals, species, and so on, in which the structure can be found. See also → redundancy and Fig. 11.

Syllogistics. Syllogistics, from the Greek,συλλέγω(= compute), is a theory of deductive inference systematically founded by Aristotle. The conclusion is inferred from two premisses, whence the possible basic forms of the premisses can be (a) universal affirmative, (b) universal negative, (c) particular affirmative, or (d) particular negative.

Synthetic theory. This term denotes the evolutionary theories generally accepted by science today. It depends essentially on the selection theory of Charles Darwin, but supported in addition by the laws of mutation theory (Neo-Darwinism) which became known after Darwin, molecular genetics and population dynamics.

Taxis. A directed movement in space is denoted as taxis, in which the organism, without trial and error, is set on the most favourable course. Here the size of the angle, in which the animal turns to the stimulus, is independent of that which the direction of the applied stimulus forms with the longitudinal axis of the body. Many → instinctive movements (perhaps the egg-rolling movement of birds) are closely linked with taxes.

Teleology. The question of the cause of numerous purpose-directed processes in living things was often explained by philosophically inclined biologists by the "immanent" principle of teleology. Since the assumption of supernatural properties has no value for real science, and attachment of living organisms to goals is manifest, we rather speak of teleonomy today. This is to point out that even purposes in living events are amenable to scientific analysis. The two expressions are distinguished perhaps like astronomy and astrology. Dialectical materialism restricts teleology to the sphere of human affairs.

Teleonomy. → teleology.

Topic. The Topic is a part of the Aristotelian Organon, dealing with the doctrine of probable inference. It "pursues the task of finding a method by which we can form conclusions on every set problem from probable postulates, without . . . falling into contradictions" (Aristotle).

Trans-specific evolution. The phylogenetic process allows classification in the evolutionary process within a species and those of the genera up to kingdoms. The latter are denoted as trans-specific evolution. In these, e.g. the phenomena of → orthogenesis, the formation of → types, the non-reversibility of evolutive processes, parallelism in the development of different groups, and so on, are involved. Causal analysis of trans-specific evolution is faced with the problem of elucidating the development laws of branching in the ancestral tree and classification in the development of large systematic units.

Type. The type of a natural relationship group is the totality of the characterising → homologous features. The morphological type is thus made up from the homologa and their trends, their positional structure, metamorphoses and coincidences. We therefore cannot readily represent it diagrammatically, since it is a dynamic time form. It is none the less as real as the structures forming it.

Vitalism. Historically, vitalism has arisen out of those philosophical theories of life that believed they could derive a non-material life principle. The new vitalism (Hans Driesch) arose as a reaction to the simple mechanistic → materialism of the 19th century and has manoeuvred itself by this contrast into an equally untenable position. For the assumed life force, "élan vital", explains the living as little as does an "élan locomotif" the function of a steam engine (Julian Huxley).

World view apparatus. → hypothetical realism.

REFERENCES

Modern editions of English classics and standard English translations of foreign originals are not given here.

Abderhalden, E. (1946): Lehrbuch der Physiologie. Wien: Urban and Schwarzenberg.

Albert, H. (1968): Traktat über kritische Vernunft. Tübingen: Mohr.

Allen, T. (1972): The marvels of animal behavior. New York: National Geogr. Soc.

Aster, E. v. (1975): Geschichte der Philosophie. Stuttgart: Kröner.

Augustinus, A. (428): De civitate Dei.

Baerends, G., K. Brill and P. Bult (1965): Versuche zur Analyse einer erlernten Reizsituation bei einem Schweinsaffen. Z. Tierpsychol. (22): 394-411.

Baltzer, F. (1955): Finalisme et physicisme. Actes Soc. Helvétique Sci. Naturelles (135): 92-99.

Bavelas, A. (1957): Group size, interaction and structural environment. 4th Conf. on Group Proc. New York: Jos. Macy Jr. Found.

Bavink, B. (1930[4]): Ergebnisse und Probleme der Naturwissenschaften. Eine Einführung in die heutige Naturphilosophie. Leipzig: Hirzel.

Bayes, T. (1763): An essay towards solving a problem in the Doctrine of Chances. Phil. Trans. Roy. Soc., Vol. 53.

Berger, P. (1977): Einladung zur Soziologie. München: Deutscher Taschenbuch Verglag.

Berger, P. and T. Luckmann (1966): The social construction of reality. New York: Doubleday.

Bertalanffy, L. v. (1955): An essay on the relativity of categories. Philos. of Science (22): 243-263.

Bertalanffy, L. v. (1968): General system theory. Foundation, development, application. New York: Braziller.

Black, M. (1954): Problems of analysis. London: Routledge and Kegan.

Bolzano, B. (1929-1931): Wissenschaftslehre I-IV. Leipzig: Meiner.

Bourne, P. (1969): The psychology and physiology of stress. New York: Academic Press.

Bresch, C. and R. Hausmann (1972): Klassische und molekulare Genetik. Heidelberg-New York: Springer.

Brickenkamp, R. (1975): Handbuch psychologischer und pädagogischer Tests. Göttingen-Toronto-Zürich: Verlag für Psychologie.

Brun, E. (1912): Zur Psychologie der Künstlichen Allianzkolonien bei den Ameisen. Biol. Zentralbl. (32): 308-322.

Brunswik, E. (1934): Wahrnehmung und Gegenstandswelt. Psychologie vom Gegenstand her. Leipzig-Wien: Deuticke.

Brunswik, E. (1939): Probability as a determiner of rat behaviour. J. exp. Psychol. (25): 175-197.

Brunswik, E. (1955): “Ratiomorphic” models of perception and thinking. Acta psychol. (11): 108-109.

Brunswik, E. (1957): Scope and aspects of the cognitive problem. In J. Bruner et al. (Ed.): Contemporary Approaches to cognition. Cambridge: Harvard University Press.

Calvin, M. (1969): Chemical evolution, molecular evolution towards the origin of living systems on the earth and elsewhere. Oxford: Clarendon Press.

Campbell, D. (1959): Methodological suggestions from a comparative psychology of knowledge processes. Inquiry (2): 152-182.

Campbell, D. (1966): Pattern matching as an essential in distal knowing. New York: Holt, Rinehart and Winston.

Campbell, D. (1974): Evolutionary epistemology. In: P. Schlipp (Ed.), 1974: The library of living philosophers. Vol. 14 I. and II.: The philosophy of Karl Popper, Vol. I: 413-463. Lasalle: Open Court.

Campbell, D. (1974a): "Downward causation" in hierarchically organised biological systems. In: F. Ayala and Th. Dobzhansky (Eds.), Studies in the philosophy of biology. London: Macmillan.

Carnap, R. (1945): The two concepts of probability. Philos. and Phenom. Res. (5): 513-532.

Carnap, R. (1952): The continuum of inductive methods. University of Chicago Press.

Carnap, R. (1967[2]): Logical foundations of probability. Chicago University Press.

Carnap, R. (1962): The aim of inductive logic. In: E. Nagel, P. Suppes and A. Tarskii (Eds.): Logic, Methodology and Philosophy of Science. Proceedings of the 1960 International Congress, Stanford (Calif.): 303-318.

Carnap, R. (1966): Philosophical foundations of physics. New York and London: Basic Books.

Carnap, R. and R. Jeffrey (1971): Studies in inductive logic and probability. Vol. I. Berkeley-Los Angeles-London.

Ceram, C. (1949): Götter, Gräber und Gelehrte. Roman der Archäologie. Reinbek: Rowohlt.

Chomsky, N. (1968): Language and Mind. New York.

Chomsky, N. (1971): Problems of Knowledge and Freedom. New York, Pantheon Books.

CIBA Foundation Symposia, (Ed.), (1963): Man and his future. Amsterdam: Excerpta medica.

Constable, G. (1973): The Neanderthals (from: The Emergence of Man). Time-Life.

Corti, W., (Ed.) (1976): The philosophy of William James. Hamburg: Meiner.

Craig, W. (1918): Appetites and aversions as constituents of instincts. Biol. Bull. Woods Hole (34): 91-107.

Crosby, E., C. T. Humphrey and E. Lauer (1962): Correlative anatomy of the nervous system. New York: Macmillan.

De Bono, E. (1969): The mechanism of Mind.

Deglin, W. (1976): Unsere zwei Gehirne. UNESCO-Kurier; 17, (1): 4-32.

Descartes, R. (1641): Méditations touchant la philosophie première.

Dessauer, F. (1958): Naturwissenschaftliches Erkennen. Frankfurt/M.: Knecht.

Diemer, A. and I. Frenzel, (Eds.), (1967): Philosophie. In: Fischer Lexikon, Bd. 11. Frankfurt/M.: Fischer.

Dilthey, W. (1933): Einleitung in die Geisteswissenschaften. Stuttgart: Teubner.

Ditfurth, H. v. (1976): Der Geist fiel nicht vom Himmel. Die Evolution unseres Bewusstseins. Hamburg: Hoffmann and Campe.

Dobzhansky, Th. (1951): Genetics and the origin of species. New York: Columbia University Press.

Döhl, J. (1966): Manipulierfähigkeit und "einsichtiges" Verhalten eines Schimpansen bei komplizierten Handlungsketten. Z. Tierpsychol. (23): 77-113.

Dörner, D. (1967): Problemlösen als Informationsverarbeitung. Stuttgart-Berlin-Köln-Mainz: Kohlammer.

Dörner, D. (1975): Wie Menschen eine Welt verbessern wollten und sie dabei zerstörten. Bild d. Wissensch. (2): 48-53.

Dörner, D. and F. Reither (1978): Über das Problemlösen in sehr komplexen Realitätsbereichen. Zeitschr. f. experimentelle und angewandte Psychologie, 25 (4): 527-551.

Driesch, H. (1909): Philosophie des Organischen (2 Bde). Leipzig: Engelmann.

Duncker, H. (1978): Das Denken in komplexen Zusammenhängen und die Fähigkeit zu kreativem Handeln. Jahresbericht d. Studienstiftung d. deutschen Volkes 1977. Bonn: Studienstiftung: 26-46.

Durant, W. (1953): The pleasures of philosophy. An attempt at a consistent philosophy of life. New York: Simon and Schuster.

Durant, W. and A. Durant (1957): The story of civilisation. 10 vol., New York: Simon and Schuster.

Eccles, J., (Ed.), (1966): Brain and conscious experience. Heidelberg: Springer.

Eccles, J. (1973): The understanding of the brain. New York: McGraw Hill.

Edey, M. (1972): The Missing Link (from: The Emergence of Man). Time-Life.

Ehrenfels, Chr. v. (1890): Über Gestaltsqualitäten. Vierteljahresschrift wissensch. Philosophie (14): 249-292.

Eibl-Eibesfeldt, I. (1970): Liebe und Hass. Zur Naturgeschichte elementarer Verhaltensweisen. München-Zürich: Piper.

Eibl-Eibesfeldt, I. (1975): Krieg und Frieden aus der Sicht der Verhaltensforschung. München-Zürich: Piper.

Eibl-Eibesfeldt, I. (1978^5): Grundriss der vergleichenden Verhaltensforschung. München-Zürich: Piper.

Eigen, M. (1971): Self-organization of matter and the evolution of biological macromolecules. Naturwiss. (58): 465-522.

Eigen, M. (1976): Wie entsteht Information? Ber. Bunsenges. physik. chem. (80): 1059-1081.

Eigen, M. and R. Winkler (1973/74): Ludus vitalis. In: H. v. Ditfurth (Ed.): Mannheimer Forum 73/74: 53-140.

Eigen, M. and R. Winkler (1975): Das Spiel. Naturgesetze steuern den Zufall. München-Zürich: Piper.

Eigen, M. and P. Schuster (1977): The hypercycle. A principle of self-organization. Naturwiss. (64): 451-565.

Einstein, A. (1949): The world as I see it. New York — London — Toronto.

Einstein, A. and M. Born (1969): Briefwechsel 1916-1955. München: Nyphenburger.

Eisler, R. (1930): Kant-Lexikon. Hildesheim-New York (1972): Olms.

Escher, M. (1975): Graphik und Zeichnungen. München: Moos.

Feyerabend, P. (1970): Against method. An anarchistic theory of knowledge. In: Minnesota Studies in the Philosophy of Science, Bd. IV. Winokur; Minneapolis.

Feyerabend, P. (1978): Der wissenschaftstheoretische Realismus und die Autorität der Wissenschaften. Braunschweig-Wiesbaden: Vieweg.

Finetti, B. de (1937): La prévision: ses lois logiques, ses sources subjectives. Annales de l'Inst. Poincaré (7): 93-158.

Finetti, B. de (1970): Teoria delle Probabilità (2 vols). Turin.

Foppa, K. (1964): Probabilistische Lernmodelle. In: Bergius, R., (Ed.): Handbuch der Psychologie, Bd. 1,2. Halbband: 617-640. Göttingen: Hogrefe.

Foppa, K. (1965): Lernen, Gedächtnis, Verhalten. Ergebnisse und Probleme der Lernpsychologie. Köln: Kiepenhauer and Witsch.

Forrester, J. (1971): Behavior of social systems. In: P. Weiss, (Ed.): Hierarchically organized systems in theory and practice: 81-122. New York: Hafner.

Frege, G. (1879): Begriffsschrift. Darmstadt (1971): Wiss. Buchgesellschaft.

Freud, S. (1940): Vorlesungen zur Einführung in die Psychoanalyse. Frankfurt/M. (1977): S. Fischer.

Freud, S. (1958): Der Witz und seine Beziehungen zum Unbewussten. Frankfurt/M.: Fischer.

Friedrich, H., (Ed.), (1968): Mensch und Tier. Ausdrucksformen des Lebendigen. München: Deutscher Taschenbuch Verlag.

Furth, H. (1969): Piaget and Knowledge. Theoretical foundations. Prentice-Hall, Englewood Cliffs, N.J.

Gabriel, G., (Ed.), (1971): Gottlob Freges Schriften zur Logik und Sprachpilosophie. Hamburg: Meiner.

Galbraith, J. (1970): The new industrial state. Boston: Houghton.

Gazzaniga, M. (1970): The bisected brain. New York: Appleton-Century-Crofts.

Gehlen, A. (1940): Der Mensch. Seine Natur und seine Stellung in der Welt. Berlin: Junker and Dünnhaupt.

Goede, K. and F. Klix (1972): Lernabhängige Strategien der Merkmalsgewinnung und der Klassenbildung beim Menschen. In: F. Klix, W. Krause, and H. Sydnow, (Eds.): Kybernetik-Forschung, Zeichenerkennung und Klassifizierungsprozesse in biologischen und technischen Systemen. Berlin.

Goethe, J. v. (1790): Morphologische Schriften. Weimar: Böhlau.

Goldscheider, P. and H. Zemanek (1971): Computor; Werkzeug der Information. Berlin-Heidelberg-New York: Springer.

Grant, D., H. Hake and J. Hornseth (1951): Acquisition and extinction of a verbal conditioned response with differing percentages of reinforcements. J. exp. Psychol. (42): 1-5.

Grant, D. and L. Schipper (1952): The acquisition and extinction of conditioned eyelid responses as a function of the percentage of fixed-ratio random reinforcements. J. exp. Psychol. (43): 313-320.

Gregory, R. (1966): Eye and Brain. London.

Gregory, W. (1951): Evolution emerging. A survey of changing patterns from primeval life to man. New York: Macmillan.

Grelling, K. (1935): Wahrscheinlichkeit von Hypothesen. Erkenntnis (5): 168-170.

Grzimek, B., (Ed.), (1968): Grzimeks Tierleben. Enzyklopädie des Tierreiches. München-Zürich: Kindler.

Hacking, I. (1965): The logic of statistical inference. Cambridge: Cambridge University Press.

Harris, N. (1970): Die Ideologien in der Gesellschaft. Eine Untersuchung über Entstehung, Wesen und Wirkung. München: Beck.

Hartmann, N. (1951): Telelogisches Denken. Berlin: De Gruyter.

Hartmann, N. (1964[3]): Der Aufbau der realen Welt. Berlin: De Gruyter.

Hartshore, C. and P. Weiss, (Eds.), (1931-58): Collected papers of Charles Sanders Peirce. Cambridge: Harvard University Press.

Hassenstein, B. (1951): Goethes Morphologie als selbstkritische Wissenschaft und die heutige Gültigkeit ihrer Ergebnisse. Neue Folge d. Jahrb. d. Goethe-Gesellschaft, (12): 333-357.

Hassenstein, B. (1954): Abbildende Begriffe. In: Verh. dtsch. Zool. Ges. 1954: 197-202.

Hassenstein, B. (1958): Prinzipien der vergleichenden Anatomie bei Geoffroy Saint-Hillaire, Cuvier und Goethe. Act. Coll. int. Strasbourg. Publ. Fac. lettr. (137): 155-168.

Hassenstein, B. (1965): Biologische Kybernetik. Heidelberg: Quelle and Meyer.

Hassenstein, B. (1969): Biologie des Lernens. In: Der Lernprozess. Willmann-Institut: 107-136. Freiburg: Herder.

Hassenstein, B. (1973): Verhaltensbiologie des Kindes. München-Zürich: Piper.

Hassenstein, B. (1974): Lern- und Spielverhalten. In: K. Immelmann, (Ed.): Verhaltensforschung; Grzimeks Tierleben, Ergänzungsband. München-Zürich: Kindler.

Hassenstein, B. (1976): Injunktion. In: J. Ritter and K. Gründer (Eds.): Historisches Wörterbuch der Philosophie, Bd. 4: 367. Basel-Stuttgart: Schwabe.

Hayek, F. v. (1952): The sensory order. An inquiry into the foundations of theoretical psychology. London: Routledge and Kegan.

Hayek, F. v. (1979[2]): Missbrauch und Verfall der Vernunft. Ein Fragment. Salzburg: Neugebauer.

Hegel, G. (1806): Phänomenologie des Geistes. In: G. Hegel: Ges. Werke, Bd. 9. Hamburg (1979): Meiner.

Heinroth, K. (1974): Die Geschichte der Verhaltensforschung. In: K. Immelmann, (Ed.): Verhaltensforschung; Grzimeks Tierleben, Ergänzungsband. München-Zürich: Kindler.

Heisenberg, W. (1966): Das Naturbild der heutigen Physik. Reinbek: Rowohlt.

Heisenberg, W. (1969): Der Teil und das Ganze. München-Zürich: Piper.

Heisenberg, W. (1976[3]): Schritte über Grenzen. Gesammelte Reden und Aufsätze. München: Piper.

Hemleben, J. (1964): Ernst Haeckel in Selbstzeugnissen und Bilddokumenten. Reinbek: Rowohlt.

Hemleben, J. (1969): Galileo Galilei in Selbstzeugnissen und Bilddokumenten. Reinbek: Rowohlt.

Hempel, C. (1945): Studies in the Logic of Confirmation (I), Mind (54): 1-12, 97-121.

Hermes, H. (1961): Aufzählbarkeit, Entscheidbarkeit, Berechenbarkeit. Berlin-Göttingen-Heilelberg: Springer.

Herrmann, T., P. Hofstätter, H. Huber and F. Weinert, (Eds.), (1977): Handbuch psychologischer Grundbegriffe. München: Kösel.

Hess, E. (1959): Imprinting. Science (130): 133-141.

Hess, E. (1975): Prägung. Die Frühkindliche Entwicklung von Verhaltensmustern von Tier und Mensch. München-Zürich: Kindler.

Hochstetter, F. (1945[19]): Toldts Anatomischer Atlas. Wien: Urban und Schwarzenberg.

Hofstätter, P. (1972): Psychologie. In: Fischer Lexikon, Bd. 6. Frankfurt/M.: Fischer.

Holst, D. v. (1969): Sozialer Stress bei Tupajas (Tupaja belangeri). Die Aktivierung des sympathischen Nervensystems und ihre Beziehung zu hormonal auglelösten ethologischen und physiologischen Veränderungen. Z. Vergl. Physiol. (63): 1-58.

Holst, E. v. (1969): Zur Verhaltensphysiologie bei Tier und Mensch. Gesammelte Abhandlungen. München-Zürich: Piper.

Holst, E. v. and H. Mittelstaedt (1950): Das Reafferenz-Prinzip. Naturwiss. (37): 464-476.

Holton, G. (1979): Einstein's model for constructing a scientific theory. In: P. Aichelburg and R. Sexl (Eds.): Albert Einstein — His influence on physics, philosophy and politics: 109-136. Braunschweig-Wiesbaden: Vieweg.

Hörz, H. and C. Nowinski (Eds.), (1979): Gesetz-Entwicklung-Information. Zum Verhältnis von philosophischer und biologischer Entwicklungstheorie. Berlin: Akademie-Verlag.

Hovland, C. (1952): A communication analysis of concept learning. Psychol. Rev. (59): 461-472.

Hovland, C. and W. Weiss (1953): Transmission of information concerning concepts through positive and negative instances. J. exp. Psychol. (45): 175-182.

Hume, D. (1748): An Inquiry Concerning Human Understanding.

Humphreys, L. (1939): Generalization as a function of method of reinforcement. J. expl. Psychol. (25): 361-372.

Hunt, E. (1962): Concept learning. New York-London: Macmillan.

Huxley, A. (1966): Brave new world revisited. London: Chatto and Windus.

Huxley J. (1929): Biology of the human race. In. H. Wells, J. Huxley and G. Wells, (Eds.): The science of life. New York.

Huxley, J. (1942): Evolution, the modern synthesis. New York: Harper and Row.
Huxley, J. (1954): Evolutionary Humanism. Melbourne: Australian Inst. of Intern. Affairs.

Jeffreys, H. (1939): Theory of probability. Oxford: Oxford University Press.
Joerger, K. (1976): Einführung in die Lernpsychologie. Freiburg-Basel-Wien: Herder.
Jouvenel, B. de (1968): Arcadie. Essais sur le mieux-vivre. Paris: Futuribles.
Jung, C. (1954): Von den Wurzeln des Bewusstseins. Zürich-Stuttgart: Rascher.

Kant, I. (1755): Allgemeine Naturgeschichte und Theorie des Himmels, oder Versuch von der Verfassung und dem mechanischen Ursprunge des ganzen Weltgebäudes nach Newtonischen Grundsätzen abgehandelt. In I. Kant, Werkausgabe; Bd. I. Frankfurt/M. (1977): Suhrkamp.
Kant, I. (1770): Von der Form der Sinnes- und Verstandeswelt und ihren Gründen. In: I. Kant, Werkausgabe Bd. V. Frankfurt/M. (1977): Suhrkamp.
Kant, I. (1781):Kritik der reinen Vernunft. In: I. Kant, Werkausgabe, Bd. III and IV. Frankfurt/M. (1977): Suhrkamp.
Kant, I. (1783): Prolegomena zu einer jeden künftigen Metaphysik, die als Wissenschaft wird auftreten können. In: I. Kant, Werkausgabe, Bd. V. Frankfurt/M. (1977): Suhrkamp.
Kant, I. (1784): Idee zu einer allgemeinen Geschichte in weltbürgerlicher Absicht. In: I. Kant, Werkausgabe, Bd. XI. Frankfurt/M. (1977:) Suhrkamp.
Kant, I. (1788): Uber den Gebrauch teleologischer Prinzipien in der Philosophie. In: I. Kant, Werkausgabe, Bd. IX. Frankfurt/M. (1977): Suhrkamp.
Kant, I. (1790): Kritik der Urteilskraft. In: I. Kant, Werkausgabe, Bd. X. Frankfurt/M. (1977): Suhrkamp.
Kant, I. (1790a): Über eine Entdeckung, nach der alle neue Kritik der reinen Vernunft durch eine ältere entbehrlich gemacht werden soll. In: I. Kant, Werkausgabe, Bd. V. Frankfurt/M. (1977): Suhrkamp.
Kant, I. (1804): Welches sind die wirklichen Fortschritte, die Metaphysik seit Leibnizens und Wolff's Zeiten in Deutschland gemacht hat? In: I. Kant, Werkausgabe, Bd. VI. Frankfurt/M. (1977): Suhrkamp.
Kaspar, R. (1977): Der Typus — Idee und Realität. Acta biotheoretica (26), 3: 181-195.
Kaspar, R. (1978): Die Geschichtlichkeit lebendiger Ordnung. Biologie in unserer Zeit (2): 42-47.
Kaspar, R. (1980): Die Evolution erkenntnisgewinnender Mechanismen. Biologie in unserer Zeit. (1):17-22.
Kaspar, R. (1980a): Kritische Anmerkungen zum panpsychistischen Identismus von Bernhard Rensch. Phil. Nat.
Kaspar, R. (1980b): Naturgesetz, Kausalität und Induktion. Ein Beitrag zur Theoretischen Biologie. Acta biotheoretica. (29): 129-149.
Kaulbach, F. (1968): Philosophie der Beschreibung. Köln-Graz: Böhlau.
Kernig, C., (Ed.), (1968): Sowjetsystem und demokratische Gesellschaft. Eine vergleichende Enzyklopädie. Bd. 2. Freiburg-Basel-Wien: Herder.
Keynes, J. (1921): A treatise on probability. London-New York: Macmillan.
Klement, H. -W.. (Ed.), (1975): Bewusstsein; Ein Zentralproblem der Wissenschaften. Baden-Baden: Agis.
Klix, F. (1976): Information und Verhalten. Kybernetische Aspekte der organismischen Informationsverarbeitung. Bern-Stuttgart-Wien: Huber.
Klix, F. and K. Goede (1968): Struktur- und Komponentenanalyse von Problemlösungsprozessen. Zeitschrift für Psychologie (174).
Klös, H. and U. Klös (1968): Gänseverwandte. In: B. Grzimeks Tierleben, Bd. VII: 269-275.

Kluge, F. (1967[20]): Etymologisches Wörterbuch der deutschen Sprache. Berlin: De Gruyter.
Koch, H. (1973): Der Sozialdarwinismus. Seine Genese und sein Einfluss auf das imperialistische Denken. München: Beck.
Koehler, O. (1941): Vom Erlernen unbenannter Anzahlen bei Vögeln. Naturwiss. (29): 201-218.
Koenig, O. (1970): Kultur und Verhaltensforschung. Einführung in die Kulturethologie. München: Deutscher Taschenbuch Verlag.
Koenig, O. (1975): Urmotiv Auge. München-Zürich: Piper.
Koestler, A. (1966): The Act of Creation. London.
Koffka, K. (1950): Principles of Gestalt psychology. London-New York: Harcourt.
Köhler, W. (1921): Intelligenzprüfungen bei Menschenaffen. Berlin: Springer.
Köhler, W. (1971): Die Aufgaben der Gestaltpsychologie. Berlin-New York: De Gruyter.
Kolmogorow, A. (1933): Grundbegriffe der Wahrscheinlichkeits-Rechnung. Berlin: Springer.
Krause, W. (1970): Untersuchungen zur Komponentenanalyse in einfachen Problemlösungsprozessen. Zeitschr. f. Psychol. (177): 199-249.
Kuhn, Th. (1962): The Structure of Scientific Revolutions. Chicago University Press.
Kühnelt, W. (1961): Soil biology; with special reference to the animal kingdom. London: Faber and Faber.
Kummer, B. (1959): Bauprinzipien des Säugerskelettes. Stuttgart: Thieme.
Kurten, B. (1974): Die Welt der Dinosaurier. Frankfurt/M.: Fischer.
Kutschera, F. v. (1972): Wissenschafts-Theorie, I und II; Grundzüge einer allgemeinen Methodologie der empirischen Wissenschaften. München: Fink.
Kuyten, P. (1962): Verhaltensbeobachtungen an der Raupe des Kaiseratlas. Entomol. Z. (72): 203-207.

Lack, D. (1943): The life of the robin. Cambridge: Cambridge University Press.
Lamarck, J. (1809): Philosophie zoologique.
Lamettrie, J. de (1747): L'homme machine.
Laplace, P. de (1796): Exposition du système du monde. Paris (1835): Bachelier.
Laplace, P. de (1812): Théorie analytique des probabilités. Paris: Bachelier.
Lawick-Goodall, J. van (1967): My Friends the Wild Chimpanzees. Nat. Geogr. Soc.
Leibniz, G. v. (1704): Neue Abhandlungen über den menschlichen Verstand. Frankfurt/M. (1961): Insel.
Leibniz, G. v. (1710): Essais de théodicée sur la bonté de dieu, la liberté de l'homme et l'origine du mal.
Lenneberg, E. (1967): Biological foundations of language. Wiley, New York.
Levy-Agresti, J. and R. Sperry (1968): Differential percentual capacities in major and minor hemispheres. Proc. Nat. Acad. Sci. U.S. (61): 1151.
Locke, J. (1690): An Essay Concerning Human Understanding.
Lorenz, K. (1941): Kants Lehre vom Apriorischen im Lichte gegenwärtiger Biologie, Blätter für Deutsche Philosophie (15): 94-125.
Lorenz, K. (1943): Die angeborenen Formen möglicher Erfahrung, Z. Tierpsychol. (5): 235-409.
Lorenz, K. (1954): Morphology and behavior patterns in allied species. New York: 1st Conf. on Group Proc. Jos. Macy Jr. Found: 168-220.
Lorenz, K. (1959): Gestaltwahrnehmung als Quelle wissenschaftlicher Erkenntnis. Zeitschr. f. exp. u. angewandte Psychol. (4): 118-165.
Lorenz, K. (1963): Das sogenannte Böse. Zur Naturgeschichte der Aggression. Wien: Borotha-Schöler.
Lorenz, K. (1965): Über tierisches und menschliches Verhalten. Aus dem Werdegang der Verhaltenslehre, (2 Bde.). München-Zürich: Piper.

Lorenz, K. (1965a): Darwin hat recht gesehen. Pfullingen: Neske.
Lorenz, K. (1971): Knowledge, beliefs and freedom. In: P. Weiss, (Ed): Hierarchically organized systems in theory and practice. New York: Hafner.
Lorenz, K. (1973): Die Rückseite des Spiegels. Versuch einer Naturgeschichte menschlichen Erkennens. München-Zürich: Piper.
Lorenz, K. (1974a): Analogy as a source of knowledge. In: Les Prix Nobel en 1973. The Nobel Foundation 1974: 176-195.
Lorenz, K. (1974b): Das wirklich Böse. Involutionstendenzen der Kultur. In: O. Schatz: Was wird aus dem Menschen? Graz-Wien-Köln: Styria.
Lorenz, K. (1974c): Die acht Todsünden der zivilisierten Menschheit. München-Zürich: Piper.
Lorenz, K. (1978): Vergleichende Verhaltensforschung. Grundlagen der Ethologie. Wien-New York: Springer.
Lovins, A. (1978): Sanfte Energie — Das Programm für die energie- und industriepolitische Umrüstung unserer Gesellschaft. Reinbek: Rowohlt.
Lukasiewicz, J. (1935): Zur Geschichte der Aussagenlogik. Erkenntnis (5): 112.
Lüscher, E. (1978): Pipers Buch der modernen Physik. München-Zürich: Piper.
Lyell, CH. (1875): Principles of geology. London: Murray.

Mach, E. (1905): Erkenntnis and Irrtum. Leipzig: Barth.
Mach, E. (1910): Die Leitgedanken meiner naturwissenschaftlichen Erkenntnislehre und ihre Aufnahme durch die Zeitgenossen. Physik.Zeit. (11): 599-606.
Mach, E. (1921^8): Die Mechanik in ihrer Entwicklung. Leipzig: Barth.
March, A. (1948): Natur und Erkenntnis. Die Welt in der Konstruktion des heutigen Physikers. Wien: Springer.
Marfeld, A. (1973): Kybernetik des Gehirns. Ein Kompendium der Grundlagenforschung. Reinbeck: Rowohlt.
Marx, K. und F. Engels (1846): Die deutsche Ideologie. In: K. Marx, F. Engels, Ausgewählte Werke in 6 Bänden. Berlin/Ost (1977): Dietz.
Mason, S. (1974): Geschichte der Naturwissenschaft. Stuttgart: Kröner.
Mayr, E. (1963): Animal Species and Evolution. Belknap Press, Cambridge (Mass.).
Mayr, E. (1970): Population, species and evolution. Cambridge: Belknap, Harvard University Press.
Metzger, W. (1963^3): Psychologie, Die Entwicklung ihrer Grundlagen seit der Einführung des Experiments. Darmstadt: Steinkopff.
Mill, J. (1872^8): System of Logic.
Mohr, H. (1967): Wissenschaft und menschliche Erkenntnis, Freiburg: Rombach.
Monod, J. (1959): Biosynthese eines Enzyms. Angewandte Chemie (71): 685-691.
Monod, J. (1970): Le hasard et la nécessité. Edition du Seuil, Paris.

Narr, K. (1961): Urgeschichte der Kultur, Stuttgart: Kröner.
Neumann, E. (1974^2): Geist und Psyche. Ursprungsgeschichte des Bewusstseins. München-Zürich: Kindler.
Norman, J. and F. Fraser (1960): Giant Fishes, Whales and Dolphins. Putnam, London.

Oeser, E. (1971): Kepler. Die Entstehung der modernen Wissenschaft. Göttingen: Musterschmidt.
Oeser, E. (1974): System, Klassifikation, Evolution, Wien-Stuttgart: Braumüller.
Oeser, E. (1976): Wissenschaft und Information. Systematische Grundlagen einer Theorie der Wissenschaftsentwicklung (3 vols.). Wien-München: Oldenbourg.
Oeser, E. (1979): Wissenschaftstheorie als Rekonstruktion der Wissenchaftsgeschichte, Fallstudien zu einer Theorie der Wissenschaftsentwicklung. Band I. Metrisierung, Hypothesenbildung, Theoriendynamik. Wien-München: Oldenbourg.
Orwell, G. (1945): Animal farm. A fairy story. Aylesbury: Hunt Barnard Printing.

Orwell, G. (1949): Nineteen Eighty-Four.
Osche, G. (1972): Evolution: Grundlagen-Erkenntnisse-Entwicklungen der Abstammungslehre. Freiburg-Basel-Wien: Herder.
Ostwald, W. (1898): Das physikalisch-chemische Institut der Universität Leipzig und die Feier seiner Eröffnung. Leipzig.

Pascal, B. (1645): De l'esprit géométrique.
Pavlov, I. (1927): Conditioned Reflexes. Oxford.
Penfield, W. and L. Roberts (1959): Speech and brain mechanisms. Princeton University Press.
Pepper, S. (1958): The sources of value. Berkeley-Los Angeles: California University Press.
Peterson, R., G. Montfort and P Hollom (1952): A field guide to the birds of Britain and Europe. Collins, London.
Piaget, J. (1946): La formation du symbole chez l'enfant. Imitation, jeu et rève, image et représentation. Neuchâtel: Delachaux et Niestlé.
Piaget, J. (1973): Einführung in die genetische Erkenntnistheorie. Frankfurt/M.: Suhrkamp.
Piaget, J. (1967): Biologie et connaissance. Gallimard, Paris.
Pickenhain, L. (1959): Grundriss der Physiologie der Höheren Nerventätigkeit. Berlin: Volk and Gesundheit.
Pittendrigh, C. (1958): Adaption, natural selection and behaviour. In: A. Roe and G. Simpson, (Eds.): Behavior and Evolution: 390-416. Yale: Yale University Press.
Planck, M. (1965): Determinismus oder Indeterminismus? Leipzig: Barth.
Plato: Apology.
Platzeck, E. (1962-1964): Raimundus Lullus, Düsseldorf: Schwann.
Polya, G. (1966): Vom Lösen mathematischer Aufgaben. Basel-Stuttgart: Schwabe.
Ponnamperuma, C. (1972): The origins of life. London: Thames and Hudson.
Popper, K. (1957): The propensity interpretation of the calculus of probability and the quantum theory. In: S. Körner and M. Price (Eds.): Observation and Interpretation. Proceedings of the 9th Symposium of the Colston Research Society: 65-70. New York-London: Butterworths Scientific Publications.
Popper, K. (1959): The Logic of Scientific Discovery. Hutchinson, London.
Popper, K. (1972): Objective Knowledge. Clarendon Press, Oxford.
Popper, K. (1975): The rationality of scientific revolutions. In: R. Harré, (Ed.): Problems of scientific revolution; progress and obstacles to progress in the sciences. The Herbert Spencer lectures 1973. Oxford: Clarendon Press: 72-101.
Premack, D. (1971): Language in Chimpanzee? Science (172): 808-822.
Prideaux, T. (1973): Cro-Magnon-Man (from: The Emergence of Man). Time-Life.

Ramsey, F. (1931): The foundations of mathematics, and other logical essays. London-New York: Macmillan.
Razran, G. (1930): Conditioned responses in animals other than dogs. Psychol. Bull. (30).
Rechenberg, I. (1973): Evolutionsstrategie. Stuttgart-Bad-Cannstatt: Frommann.
Remane, A. (1971[2]): Die Grundlagen des natürlichen Systems, der vergleichenden Anatomie und Phylogenetik. Königstein/Taunus: Koeltz.
Rensch, B. (1954) Neuere Probleme der Abstammungslehre. Stuttgart: Enke.
Rensch, B. (1965): Homo sapiens. Göttingen: Vandenhoeck und Ruprecht.
Rensch, B. (1968): Biophilosophie. Stuttgart: G. Fischer.
Rensch, B. (1973): Gedächtnis, Begriffsbildung und Planhandlungen bei Tieren. Hamburg-Berlin: Parey.
Riedl, R. (1970[2]): Fauna und Flora der Adria. Hamburg-Berlin: Parey.
Riedl, R. (1975): Die Ordnung des Lebendigen. Systembedingungen der Evolution. Hamburg-Berlin: Parey.

Riedl, R. (1976): Die Strategie der Genesis. Naturgeschichte der realen Welt. München-Zürich: Piper.

Riedl, R. (1977): A systems-analytical approach to macro-evolutionary phenomena. The Quarterly Review of Biology (52): 351-370.

Riedl, R. (1978/79): Über die Biologie des Ursachen-Denkens. Ein evolutionistischer, systemtheoretischer Versuch. In: H. v. Ditfurth, (Ed.): Mannheimer Forum 78/79: 9-70.

Riedl, R. (1979): Die kopernikanischen Wenden. Auseinandersetzungen im abendländischen Weltbild. In: H. Huber und O. Schatz (Eds.): Glaube und Wissen; Bericht über das Münchner Symposium 1978. Wien-Freiburg: Herder.

Riopelle, A. (1972): Learning how animals learn. In: The marvels of animal behavior. Washington: Nat. Geographic. Soc.

Rohracher, Harald (1948): Die Einstellung zum Abstammungsproblem und zur psychophysiologischen Abhängigkeit. Wien: Dissertation.

Rohracher, Hubert (1965): Steuerung des Verhaltens durch Einstellung. In: H. Hekhausen (Ed.): Bericht über den 24. Kongress der Deutschen Gesellschaft für Psychologie. 1-9.

Rohracher, Hubert (1971[10]): Einführung in die Psychologie. Wien-München-Berlin: Urban and Schwarzenberg.

Romer, A. (1962[2]): The vertebrate body. Saunders, Philadelphia.

Rothacker, E. (1930[2]): Einleitung in die Geisteswissenschaften. Tübingen: Mohr (Siebeck).

Rousseau, J. (1762): Du contrat social, ou principes du droit politique.

Russell, B. (1946): History of Western philosophy. London: Allen and Unwin.

Russell, B. (1963): Has Man a Future?

Russell, B. (1948[20]): The Problems of Philosophy. OUP.

Russell, B. and A. Whitehead (1910-1913): Principia Mathematica I/III. Cambridge: Cambridge University Press.

Sachsse, H. (1967): Naturerkenntnis und Wirklichkeit. Braunschweig: Vieweg.

Sachsse, H. (1968): Die Erkenntnis des Lebendigen. Braunschweig: Vieweg.

Sachsse, H. (1971): Einführung in die Kybernetik; unter besonderer Berücksichtigung von technischen und biologischen Wirkungsgefügen. Braunschweig: Vieweg.

Savage, L. (1954): The Foundations of Statistics. New York.

Savage, L. (1967): Implications of personal probability for induction. Journal of Phil. (64): 593-607.

Schelling, F. (1797): Ideen zur Philosophie der Natur. In: F. Schelling, Werke, 1. Erg. Bd/ München (1959): Beck.

Schilpp, P. (1963): The philosophy of Rudolf Carnap. Lasalle (Illinois): Open Court.

Schleidt, W. (1962): Die historische Entwicklung der Begriffe "Angeborenes auslösendes Schema und "Angeborener Auslösemechanismus". Z. Tierpsychol. (19): 697-722.

Schmökel, H. (1966): Das Gilgamesch-Epos. Stuttgart: Kohlhammer.

Schrödinger, E. (1944): What is life?

Schumacher, E. (1973): Small is beautiful. A study of economics as if people mattered. London: Blond and Briggs.

Schuster, P. (1972): Vom Makromelekül zur primitiven Zelle. Die Entstehung biologischer Funktion. Chemie in unserer Zeit (6): 1-16.

Schwabl, H. (1958): Weltschöpfung. In: Paulys Realencyklopädie der klassischen Altertumswissenschaften. Suppl. Bd. 9: 1-142. Stuttgart: Druckenmüller.

Seitelberger, F. (1973): Das Bild des Menschen in der Sicht der Hirnforschung. Österr. Akademie d. Wiss.; math.-naturwiss. Kl., Sb. Abt. I,Bd. 181: 38-50.

Seitelberger, F. (1975): Gehirn und Umwelt. Österr. Ärztezeitung 30, (19): 1-10.

Selye, H. (1957): Stress beherrscht unser Leben. Düsseldorf: Econ.

Sexl, R. (1979): Irreversible Prozesse. In: Physik und Didaktik. Bamberg: Bayrischer Schulbuch-Verlag.
Shimony, A. (1971): Perception from an evolutionary point of view. J. Philosophy (68): 571-583.
Simon, H. and K. Kotovsky (1963): Human acquisition of concepts for sequential patterns. Psychol. Review 70(7): 534-546.
Simpson, G. (1963): Biology and the nature of science. Science (139): 81-88.
Simpson, G. (1964): Organisms and molecules in evolution. Science (146): 1535-1538.
Skinner, B. (1971): Beyond freedom and dignity. Knopf, New York.
Sneath, P. and R. Sokal (1973): Numerical taxonomy. The principle and practice of numerical classification. San Francisco: Freeman.
Sokal, R. and P. Sneath (1963): Principles of numerical taxonomy. San Francisco: Freeman.
Solecki, R. (1971): Shanidar; the first flower people. New York: Knopf.
Sperry, R. (1970a): Perception in the absence of the neocortical commissures. In: Perception and its disorders. Res. Publ. A.R.N.M.D. (The Association for Research in Nervous and Mental Disease) Bd. 48.
Sperry, R. (1970b): Cerebral dominance in perception. In: F. Young and D. Lindsley (Eds.): Early experience in visual information processing in perceptual and reading disorders. Washington: Nat. Acad. Sci.
Staudacher, W. (1942): Die Trennung von Himmel und Erde. Tübingen: Bölzle.
Stegmüller, W. (1954): Der Begriff des synthetischen Urteils a priori und die moderne Logik. Z. philosoph. Forschung (8): 535-563.
Stegmüller, W. (1969): Probleme und Resultate der Wissenschaftstheorie und analytischen Philosophie. Berlin-Heildelberg-New York: Springer.
Stegmüller, W. (1971): Das Problem der Induktion: Humes Herausforderung und moderne Antworten. In: H. Lenk, (Ed.): Neue Aspekte der Wissenschaftstheorie: 13-74, Braunschweig, Vieweg.
Stegmüller, W. (1973): Personelle und statistische Wahrscheinlichkeit. Heidelberg-New York: Springer.
Stegmüller, W. (1974): Das ABC der modernen Logik und Semantik. Der Begriff der Erklärung und seine Spielarten. Berlin-Heidelberg-New York: Springer.
Stegmüller, W. (1975): Hauptströmungen der Gegenwartsphilosophie (2 vols.). Stuttgart: Kröner.
Stirner, M. (1866): Der Einzige und sein Eigentum. Stuttgart: Reclam.
Stoppard, T. (1967): Rosencrantz and Guildenstern are dead. Play. London: Faber.
Störig, H. (1972): Knaurs moderne Astronomie. München-Zürich: Droemer-Knaur.
Strombach, W. (1970): Die Gesetze unseres Denkens. Eine Einführung in die Logik. München: Beck.

Teilhard de Chardin, P. (1957): Le phénomène humain. Editions du Seuil, Paris.
Tembrock, G. (1963): Grundlagen der Tierpsychologie. Berlin: Akademie-Verlag.
Thenius, E. and H. Hofer (1960): Stammesgeschichte der Säugetiere. Eine Übersicht über Tatsachen und Probleme der Evolution der Säugetiere. Berlin-Göttingen-Heidelberg: Springer.
Thorpe, W. (1963): Learning and instinct in animals. London: Methuen.
Tinbergen, N. (1951): The study of instinct. London: Oxford University Press.
Tinbergen, N. (1963): The Herring Gull's World. London: Collins.
Tinbergen, N. and D. Kuenen (1939): Über die auslösenden Reizsituationen der Sperrbewegung von jungen Drosseln (Turdus m. merula L. und T.e. ericetorum Turton).Z. Tierpsychol. (3): 37-60.

Urey, H. (1952): The planets. Chicago: University of Chicago Press.

Vogel, S. (1975): Mutualismus und Parasitismus in der Nützung von Pollenträgern. Verh.Dtsch.Zool.Ges.; 102-110.
Vogel, S. (1978): Evolutionary shifts from reward to deception in pollen flowers. In: A. J. Richards (Ed.); The pollination of flowers by insects. Linn. Soc. Symp. Ser. (6); 89-96.
Volkmann, P. (1913²): Einführung in das Studium der theoretischen Physik. Leipzig-Berlin: Teubner.
Vollmer, G. (1975): Evolutionäre Erkenntnistheorie. Stuttgart: Hirzel.
Voltaire, J. (1759): Candide ou l'optimisme. Paris: Miret.

Waddington, C. (1954): Evolution and epistemology. Nature (173): 880-881.
Walsh, K. (1978): Neuropsychology. Edinburgh-New York: Livingstone.
Walter, W (1951): A machine that learns. Scientific American 185 (2): 60-63.
Watson, J.B. (1925): Behaviorism.
Watson, J.D. (1977³): Molecular Biology of the Gene. London-Amsterdam-Ontario-Sidney: Benjamin.
Watzlawick, P. (1976): Wie wirklich ist die Wirklichkeit? Wahn, Täuschung, Verstehen. München-Zürich: Piper.
Weinberg, S. (1977): The first three minutes. A modern view of the origin of the universe. Basic Books, New York.
Weiss, P. (Ed.) (1971): Hierarchically organized systems in theory and practice, New York: Hafner.
Weizsäcker, C. v. (1971): Die Einheit der Natur. München: Hanser.
Weizsäcker, C. v. (1977): Der Garten des Menschlichen. Beiträge zur geschichtlichen Anthropologie. München-Wien: Hanser.
Wellek, A. (1955): Ganzheitpsychologie und Strukturtheorie. Bern: Francke.
Wentscher, E. (1921):Geschichte des Kausalproblems in der neuen Philosophie. Leipzig: Meiner.
Wertheimer, M. (1925): Drei Abhandlungen zur Gestalttheorie. Erlangen.
Wesselski, A. (1947): Deutsche Märchen vor Grimm. Wien: Rohrer.
Whewell, W. (1837): History of the Inductive Sciences, London.
Whewell, W. (1858): Novum Organon Renovatum. London: Parker and Son.
Whewell, W. (1860): On the philosophy of discovery. London: Parker and Son.
Whitehead, A. (1929): The function of reason.
Wickert, J. (1972): Albert Einstein. Reinbek: Rowohlt.
Wickler, W. (1968): Mimikry: Nachahmung und Täuschung in der Natur. München-Zürich: Kindler.
Wickler, W. und U. Seibt (1977): Vergleichende Verhaltensforschung. Hamburg: Hoffmann und Campe.
Windelband, W. (1894): Geschichte und Naturwissenschaft. Rektoratsrede an der Universität Strassburg.
Wuketits, F. (1978): Wissenschaftstheoretische Probleme der modernen Biologie. Berlin: Duncker und Humblot.

Zemanek, H. (1964): Lernende Automaten. In: K. Steinbuch (Ed.): Taschenbuch der Nachrichtenverarbeitung: 1418-1480: Berlin-Göttingen-Heidelberg: Springer.
Zemanek, H. (1968): Die Künstliche Schildkröte von Wien. Radio-Magazin mit Fernseh-Magazin (9): 275-278.
Zimmermann, W. (1953): Evolution: Die Geschichte ihrer Probleme und Erkenntnisse. Freiburg: Alber.
Zorn, R. (1952): Das Problem der Freiheit. München: Isar.

AUTHOR INDEX

SUBJECT INDEX